Dr Ruth Valerio is Global Advocacy and Influencing Director at Tearfund. An environmentalist, theologian and social activist, Ruth holds a doctorate from Kings College London, and honorary doctorates from the Universities of Winchester and Chichester. She is Canon Theologian at Rochester Cathedral and her home church is part of the 24/7 Prayer Network. She enjoys living sustainably – practising what she preaches and inspiring others as she does so – in the south of England with her family. She is the author of *L is for Lifestyle: Christian living that doesn't cost the earth* and *Just Living: Faith and community in an age of consumerism.*

All royalties from the sale of this book will be used to support Tearfund's work in the places of greatest need.

SAYING YES TO LIFE

RUTH VALERIO

First published in Great Britain in 2020

Society for Promoting Christian Knowledge
36 Causton Street
London SW1P 4ST
www.spck.org.uk

British Library Cataloguing-in-Publication Data
A catalogue record for this book is available from the British Library

ISBN 978 0 281 08377 0
eBook ISBN 978 0 281 08378 7

Typeset by The Book Guild Ltd, Leicester, UK
First printed in Great Britain by Jellyfish Print Solutions
Subsequently digitally reprinted in Great Britain

eBook by The Book Guild Ltd, Leicester, UK

Produced on paper from sustainable forests

With deep appreciation for Peter and Miranda Harris,
Susanna and Chris Naylor

Contents

Foreword

At the beginning of Lent in 2017, I travelled to Fiji to attend a meeting of the primates of the Anglican Communion in the Oceania region. Fiji is a place that has already begun to see the devastating consequences of climate change. Water levels have been rising, forcing populations to relocate, and cyclones have devastated communities. It was while I was there that one of the primates said to me, so memorably, that 'For you Europeans, climate change is a problem for the future. For us, it is a problem of everyday survival'.

As part of the global church we are called to care not just for God's creation, but for our brothers and sisters all over the world who face having their families uprooted and their livelihoods destroyed by the effects of climate change. Every single one of us has a responsibility as part of our discipleship to Jesus Christ to live a life that cares for God's world and its creatures. As Psalm 24 reads, 'The earth is the Lord's, and everything in it, the world, and all who live in it; for he founded it on the seas and established it on the waters.'

Sadly, Christians have not always given God's creation the reverence it deserves. The Old Testament offers us a picture of human beings as intimately linked to their environment, where their actions have a profound effect on the land, and they are held responsible for it, from the 'dominion' of Genesis to the covenant with Noah that holds them accountable for the blood of every creature, to the laws of Leviticus and Deuteronomy and their concern for the land. The idea of dominion has been interpreted by some Christians to mean that we can do whatever we want to the land – that it is ours to exercise our control and power over, whatever the cost. Yet this is profoundly mistaken, and fails to note the heavy responsibility laid on human beings, and their complete interdependence with creation, which means there is no space for human flourishing outside of the flourishing of the natural world.

It is of the utmost importance that we now stand in solidarity together, repenting of our sins towards our earth and committing to face our responsibilities as God's people. As people of faith, we cannot just say what we believe. We are obliged to live out the life that Christ calls us to live, to care for our neighbours, for the creatures and the creation that God has so generously given us.

We all know that sin is the state of being estranged from God. But in mistreating and abusing his creation, God's gift to us, we are also estranged. There is a need to rebuild our relationship with our planet so that we might rebuild our relationship with its creator.

Lent is a time for us to focus on Jesus' death and resurrection, and our reconciliation and atonement with God through his sacrifice. This year, I hope you might spend some time thinking about our reconciliation with God's creation as we explore the creation story of Genesis 1 together. Ruth Valerio's book is perfect for individuals and groups to think, reflect, pray and be challenged together.

The most frequent command in the Bible is 'Do not be afraid!' God isn't saying 'everything's fine, there is nothing to be afraid of' when he commands us not to fear. He is acknowledging that life is scary, and sometimes we are rightly afraid when we are confronted with such existential issues. But God is beside us, working with us in our communities and our churches, in our politics and our governments and he will not leave us to face our fears alone. This Lent is an opportunity for us to consider and respond to God's call, to live a life that is caring and restorative. We rise to the challenge with optimism and perseverance, alongside our fellow Christians from around the world, so that we might live lives that are led by Christ and shaped by the Holy Spirit, in deep reverence for God's creation.

+ + *Justin Cantuar*
Lambeth Palace, London

Acknowledgements

I could never have written *Saying Yes to Life* on my own – it is the product of many wonderful, helpful and amazing people, and I am deeply grateful to them. The time they were willing to offer reflects the dedication they feel to the issues this book covers and their desire to see the Church around the world rise up and respond.

I owe a big debt of gratitude to Tearfund, both for giving me the time and space needed to write and then for helping with the different aspects of the book, which quickly gained a life all of its own! Thank you to Nigel Harris, Lucie Woolley, Charlotte Wyatt, Naomi Henry, Ben Niblett, Jack Wakefield, Alice Philip, Hannah Swithinbank, Mari Williams, Clark Buys, Rich Gower, Sue Willsher, Miles Giljam, Naomi Foxwood, Frank Greaves, Maria Andrade, Amy Church, Levourne Passiri, Rusty Pritchard, Kossi Agbo and Sas Conradie. The Theological Committee also gave invaluable advice: thank you to Jill Garner, Jim Ingram, Mark Melluish, Harold Miller, Elaine Storkey, Stafford Carson and Rosalee Velloso Ewell. I work with truly inspiring people and it is a privilege to be part of such a Christ-centred organization, involved in bringing hope and transformation all around the world.

Folks I know through A Rocha have been hugely helpful, both in checking my theology and making sure I didn't make any ornithological or other nature-related bloomers. Thank you to Dave Bookless, Jeremy Lindsell, Hilary Marlow (who went above and beyond), Simon Marsh, Bob Sluka and Simon Stuart. Thank you also to Ruth Bancewicz, Ed Brown and Andrew Leake, and to CAFOD and Christian Aid for help with their stories.

Thank you to members of the Church of England Environment Working Group who offered invaluable feedback on the concept and draft chapters: Mark Betson, Giles Goddard, Nick Holtam, Catherine Ross, Stephen Taylor and Graham Usher. Big thanks also to Rachel Marsh for giving me so much of her time and input, and to Richard Bauckham, David Wilkinson and Jonathan Wittenberg. And of course a huge thank you to Archbishop Justin for the opportunity to write this book!

Thank you to my editor, Alison Barr, for steering me so patiently through this process, and I always feel I should say a big thank you to all my social media friends who have answered random questions, celebrated with me, and allowed me the occasional rant.

Finally, my deepest gratitude goes to my family and friends who have loved and supported me through this time. You know who you are.

The publisher and author acknowledge with thanks permission to reproduce extracts from the following:

'Oceans (Where Feet May Fail)', words and music by Matt Crocker, Joel Houston & Salomon Ligthelm. Copyright © 2012 Hillsong Music Publishing.

Psalms 1 and 8 from *Psalms Now,* 3rd edition, Leslie F. Brandt. Copyright © 1974, 1996 Concordia Publishing House, 2003. Used with permission. www.cph.org.

Song of the Shepherds, Richard Bauckham. Used by permission of the author.

The Celtic Wheel of the Year: Celtic and Christian seasonal prayers, Tess Ward. O Books, 2007. Used by permission of O Books.

'The Moon Shines Red', David Smale & Dave Withers. Copyright © 1983 Thankyou Music admin CapitolCMGPublishing.com excluding the UK & Europe which is admin by Integrity Music, part of the David C Cook family, songs@integritymusic.com.

'The Servant King', Graham Kendrick. Copyright © 1983 Thankyou Music admin CapitolCMGPublishing.com excluding the UK & Europe which is admin by Integrity Music, part of the David C Cook family, songs@integritymusic.com.

Introduction

In the beginning (Genesis 1.1–2)

> [1]In the beginning God created the heavens and the earth. [2]Now the earth was formless and empty, darkness was over the surface of the deep, and the Spirit of God was hovering over the waters.

> In the beginning, before the heavens and the earth existed, lived a god and a goddess: Apsu, the god of fresh water, and Tiamat, the goddess of salt water. Before meadowlands or reedbeds had been formed, when there were no other deities and no destinies had yet been decreed, these two mingled their waters and from those waters came younger gods. Those younger gods grew in strength and stature and became wise and mighty.

I spoke at a conference recently where I began my talk in this way. I hadn't intended to alarm people, but apparently when I opened with the line, 'In the beginning there was a god and a goddess', a delegate sitting next to a friend of mine turned to him and said in a worried whisper, 'I thought this was a Christian conference; why is the speaker saying this?' As I continued with the story that we shall hear more of below, I could feel the atmosphere change and people began to be concerned I had lost my biblical rooting!

So why start a Lent book with a strange tale of gods and goddesses?

This story forms the beginning of the Mesopotamian creation poem called *Enuma Elish*, thought to date back possibly to 1800BC (the title is taken from the opening of the poem, and means 'When on high'), and it is a fascinating account to consider in relation to Genesis 1.

Through these next six weeks of Lent we are going to read Genesis 1 together and see how the themes of this opening chapter of the Bible weave their way through the biblical story and into our lives. We will use the Days of creation to open our eyes to the world around us: to the people who live in it with their diversity, gifts and struggles, and to the other inhabitants who share our space and the environments within which we live, in all their wonder, beauty and fragility. Each chapter will look at one of the Days of creation and will focus on the various aspects of the natural world that are created on that Day. We shall see how these aspects feature in the Bible and then consider their place in our contemporary world. *Saying Yes to Life* is therefore a mix of biblical reflection and contemporary application, and through its pages we shall see churches and Christians working to bring resurrection life into many situations.[1]

As Christians, we live our lives out of the reality of the death and resurrection of Jesus Christ, and we remember his sacrifice every time we take communion throughout the year. However, Lent is a time for us to focus on these things particularly. During its forty days, we fast or give up certain things, and dedicate extra time – on our own and with others – to prepare ourselves for the events of Holy Week and Jesus' death, which led ultimately, of course, to Easter Day and the victorious resurrection. Lent thus gives us an opportunity to reflect on our wrong-doing and its impact, and to consider what practices of resurrection hope we might then take on.

You may read this book either on your own or as part of a Lent group. Each chapter ends with questions for reflection and discussion and a prayer written by a young person from one of the six human-inhabited continents of the world. It is young people who will live in the world that is being created today and, young and old, we must stand together, act together and pray together. There is a wealth of resources online at <www.spckpublishing.co.uk/saying-yes-resources>, which will help you discover more about the topics we consider in each chapter and give you further ideas for taking action. On these pages you will also find special interviews with leading experts from around the world to watch and bring into your discussions and reflections. Don't miss those!

But before we delve into Day One in Chapter One, we need to return to our story of gods and goddesses because, if we are going to appreciate Genesis 1 properly, we must spend some time understanding the wider context within which it came into being.

I am aware that thinking about the context for Genesis 1 might be a new idea for some of us reading this book. We may never have considered the origins of these creation texts and the fact that they were written in and shaped by a specific time and place in Israelite history. Yet it is key to know that these accounts have their background within the world of the ancient Near East, and that there were other stories about how the world came into being (and about a place called Eden and a flood) that originated with Israel's neighbours. These had been circulating for many years before the time it is thought that the Genesis narratives found their final form. The other stories are almost certainly older than the Genesis texts and it seems likely the writers of the biblical passages were familiar with them and used them when writing their own accounts.

Let us listen to more of the story we started at the beginning of the chapter, picking up where we left off, with the younger gods having just been begotten by Apsu and Tiamat:

> Unfortunately, as well as being strong and mighty, the young gods also became unruly and troublesome. In true fashion, as young men, they banded together as brothers and ran amok amongst the gods, dancing and partying loudly. They did this so much that they upset Tiamat's nerves, disturbed Apsu's sleep and stopped his daytime rest!
>
> Apsu decided enough was enough and he plotted to destroy the young gods in order to get his peace back. However, one of them, Ea, heard about the plot and, putting Apsu into a deep sleep, he killed him and created his own dwelling place out of Apsu's body. Tiamat was furious at what Ea had done, and she gathered around her a different set of gods and created demons – including the Hydra, the Scorpion-man and the Hairy Hero (yes really!) – in order to bring down Ea and his brothers. One particular son was called Marduk, the god of all gods, so amazing and wondrous that he could scarce be looked at. With her new husband, Kingu, Tiamat ruled and introduced a reign of terror and chaos.
>
> When Ea heard of Tiamat's fury and desire for revenge, and about the hideous army she had drawn around her and her rule

of chaos, he sent various of the gods to try to appease her, but none of them had the courage even to approach her. The gods met to discuss what could be done, and finally Marduk agreed that he would fight her, on condition they appointed him king. This they did and so Marduk set out and, in a mighty battle, captured Tiamat, the demons and her divine allies. Having slain the gods (and tied the demons to his feet), he went back to Tiamat and killed her too, in a manner far too gruesome to repeat here!

Marduk sliced Tiamat's dead body in two and formed the heavens from one half of her and the earth from the other. He created the Euphrates and Tigris rivers to flow from her eyes and from her breasts, he created the mountains. Then, from the blood of the defeated and slain Kingu, he created humankind, to serve the gods and set the gods free from having to do any work.

And thus the world and its inhabitants were created.

Taming the sea dragon

Enuma Elish, the Babylonian creation story, is based on an older Sumerian myth, which was re-appropriated when Babylon conquered the wider Mesopotamian region. The story was designed to establish the supremacy of Babylon as the ruling kingdom, with Marduk as the supreme god.

When we read this and other ancient Near Eastern stories and compare them with the texts of Genesis 1–11, particularly the narratives around the creation and the flood (though we will focus on the creation story and put the flood to one side), we find they carry a number of similarities, but also some big differences. What was going on? Had the writers of our Genesis stories simply copied this and other stories to develop their own versions or was something else at play?

Both the Genesis texts and *Enuma Elish* start from a place of chaos. The notion of 'the deep' (*tĕhôm* in the Hebrew) has resonances with the Babylonion *tamtu*, which relates to the goddess Tiamat, representing the primordial sea or ocean. Both texts begin the actual creation with the separation of heaven and earth, and both texts describe the establishing of the rhythm of day and night and the sacred seasons.

Yet here is where the similarities end and it becomes clear that one of the purposes of the opening chapters of the Bible was to challenge the society around Israel about the prevailing views on gods, the world and the status of the people who inhabit it.

As the creation stories come at the beginning of the Bible, we tend to think they were the first texts to be written. However, it is actually more likely that they are some of the later texts of the Old Testament and that, although versions of the stories may have been circulating orally for some while, they assumed their final form when the Israelites were in captivity in Babylon. This was around the time prophets such as Jeremiah and Ezekiel were speaking.

The Israelites found themselves in a situation of exile. They were strangers in a strange land, surrounded by peoples, customs and religions very different from their own. Babylon was a land of domination, where kings ruled absolutely and the rest of the population were their subjects – not surprising when their creation stories were centred around murder, double-crossing and conquest. Nothing of the Israelite understanding of justice and caring for the land was known. None of the laws of Babylon talked about looking after the widow, or the stranger, or giving back land to the dispossessed, or leaving the edges of the fields unharvested or allowing the animals to rest. So the people of Israel found themselves in a crisis of faith: disoriented and dislocated.

Into this context comes Genesis 1, a narrative written in stark contrast to the account of the Babylonians; a narrative that has stood the test of time and whose influence has resounded through the millennia. As we shall see, its message is one of hope, peace and confidence: a good God who reigns supreme has created a very good world, with people created to work with him in taking care of it and one another, not as his slaves but as his friends.

This is what we are going to explore as we go through the Days of creation together. We will look at themes of light and water; land and trees; sea creatures and birds; land animals and then, finally, the last animal to be created: humans. We will consider how all these themes weave together and what an amazing, but troubled, world we live in. We will see churches and Christians working to bring resurrection life into many situations, and we shall think about what action we can take ourselves.

In the beginning God created the heavens and the earth

Millions of Christians state this belief around the world every Sunday. It is foundational to our faith, and the topics we shall be looking at in this book are foundational to what it means to be a follower of Jesus. For too long, the theology and the practice of caring for people and planet have been sidelined in the Christian faith. It is high time to bring them into the centre and root them strongly in our churches and Christian lives. My prayer is that this book will play a small part in helping that to happen.

1
Let there be light (Genesis 1.1–5)

[3]And God said, 'Let there be light,' and there was light. [4]God saw that the light was good, and he separated the light from the darkness. [5]God called the light 'day,' and the darkness he called 'night.' And there was evening, and there was morning—the first day.

Some years ago I travelled by train to Innsbruck in the Austrian Alps. I made friends with another woman taking the same route. It was her first time outside the UK and we both revelled in the sights we saw from the train window. At one point, we turned a corner on the track and the Alps came into full view. 'Wow,' exclaimed my new-found companion, 'it must've been quite something when God coughed up that lot!'

Over the course of *Saying Yes to Life* we are going to deepen our appreciation of all that 'God coughed up' as we travel together through the unfolding story of creation. In this chapter, we start at the beginning with the creation of light. We shall see where and how light features in the Bible and will look at light today, exploring some of the extraordinary creatures who use it in extraordinary ways, and the importance of light in the form of electricity for people. This will lead into thinking about energy and issues of poverty and climate change and the urgent need for all of us to take action.

Before we do that, though, we need to take a broad look at the significance of seeing God as creator and how that relates to our understanding of salvation. When we compare the Genesis creation accounts with the Babylonian story we considered in the Introduction, there is one prime point that comes through, and that is the depiction of God.

The supreme God

In *Enuma Eelish*, as we have seen, there are many gods, demons and a goddess, fighting, squabbling, playing, messing around, and giving birth to other gods. Although Marduk emerges triumphant, there are a host of other gods around him (and presumably those other gods could in time rise up against him and put another in his place, even if the text itself could not allow for that possibility). The Genesis account will have none of this and confirms the foundational Israelite assertion called the Shema: 'Hear O Israel, the Lord our God, the Lord is one' (Deut. 6.4). There are no other gods: Yahweh alone is the one true God and the only God to be worshipped and followed. He is not a tribal god: he is the God of the whole world.

It is this God who creates the world, and creates it simply 'at his command', as Psalm 148 declares (v. 5). There is no battle for supremacy that God has to engage in: he doesn't have to fight a sea dragon or struggle against other gods and demons. There is no conflict in the Genesis account of creation: God simply speaks 'let there be light', and it comes to be.

Psalm 33.6–9 is a beautiful declaration of this truth:

By the word of the Lord the heavens were made,
  their starry host by the breath of his mouth.
He gathers the waters of the sea into jars,
  he puts the deep into storehouses.
Let all the earth fear the Lord;
  let all the people of the world revere him.
For he spoke, and it came to be;
  he commanded and it stood firm.

As we start our Lenten journey, it is right that we begin by reflecting not on ourselves or on the wider aspects of the created order, but on God, who brought all these things into being. Through *Saying Yes to Life* we shall be looking at many topics, all of which are important for us to think about and respond to. The most vital, however, is our God, 'the only one who in his very being is life and love, the uncreated one who is the source of all goodness, the God who made all things'.[1] This is the God whom we worship and whom we are committing ourselves to again over these forty days of Lent.

We are offered a more formidable depiction of this God when we look at his words to Job.[2] Chapters 38–41 are among the most stunning words of the Bible:

> Then the Lord spoke to Job out of the storm. He said:
> [. . .]
> 'Where were you when I laid the earth's foundation?
> Tell me, if you understand.
> Who marked off its dimensions? Surely you know!
> Who stretched a measuring line across it?
> On what were its footings set,
> or who laid its cornerstone—
> while the morning stars sang together
> and all the angels shouted for joy?' (Job 38.1, 4–7)

Although we cannot look at these chapters fully here, I encourage you to put this book down for a moment and read them through . . .

These words hit us with the power and might of God. He is the one who laid out the earth's foundations and marked off its dimensions; who shut up the seas behind doors, making the clouds its limits. The poem about the sea creature Leviathan in chapter 41 may be an oblique reference to Tiamat as the sea monster. However as Psalm 104.26 reminds us, for Yahweh God, Leviathan is not a hideous goddess who needs to be battled against, but simply a creature, overwhelming for us in its proportions, but for God like a pet to be sported with.

The opening verse of Genesis 1 has the word *bara*, 'to create', a word that is only used of God in the Old Testament. And yet he is not only a powerful, awe-inspiring God. In the Job passages there is a beautiful tenderness in how he watches 'when the doe bears her fawn' (Job 39.1) and then observes the young 'thrive and grow strong in the wilds' (Job 39.4). He is intensely interested and involved in all aspects of his creation, from the huge storehouses of snow and hail, to the wild donkey in the wasteland. There is even a note of comedy: one gets the impression that God finds the ostrich lovingly funny with her ungainly inability to fly and lack of good sense!

All in all, the creation narratives, both in Genesis and elsewhere in the Old Testament, show God as the ultimate creator of all that is: a God to be feared and worshipped. However, though we say that 'we believe in God, the Father almighty, creator of heaven and earth', we too often leave that statement behind and forget that God as creator is the foundation for all our other beliefs. God as creator is also the foundation for what we do. The late Prof. Dr Wangari Maathai, the founder of the Green Belt Movement which has planted over 45 million trees across Kenya, reflected on the difference it would make if we remembered creation as well as salvation. 'One is left to wonder whether conceiving of God as the origination of all that is would make people of faith recognize that they have a responsibility to be the custodians of God's creation and, in the process, their own survival'.[3]

Holding creation and salvation together

If we do neglect the doctrine of creation in our churches and in our own thinking, it is often because we have developed an understanding of salvation that does not hold salvation and creation together.

Lent is the time when we focus particularly on Jesus' death and resurrection and on the redemption he bought us. During Lent we deliberately create space to remember again the time leading up to Easter Day when Jesus travels towards Jerusalem and then goes through the events that we commemorate during Holy Week. Our lives as followers of Jesus are built on the salvation that is ours because of what Jesus did for us.

This Lent, my prayer is that *Saying Yes to Life* will help you remember the link between redemption and creation. Too often the doctrine of creation is divorced from redemption, but, as the Kenyan theologian Samson Gitau expresses it, 'There can be no redemption without creation'.[4] The Genesis creation narratives, along with other passages we have just touched on, briefly give us a glorious vision of God speaking an amazing world into being, one that is teeming with life and has God's goodness at the heart of it. The ninth century

Celtic theologian, Eriugena, said, 'It was to bring human nature back to this vision that the Incarnate Word of God descended, sweeping away the shadows of false phantasies [sic], opening the eyes of the mind, showing Himself in all things'.[5] The redemption that we anticipate eagerly as we go through Lent is not one that takes us away from creation but one that will root us more deeply in it, as we find there the light and life of God. As the Celtic spirituality scholar, Philip Newell writes, grace and nature are inextricably linked: 'The gift of grace is given to restore us to the essential well-being of the gift of nature'.[6]

The magnificent hymn of Colossians 1.15–20 makes the startling statement that Christ's blood was shed on the cross 'to reconcile to himself all things, whether things on earth or things in heaven'. Many of us are used to seeing Christ's work on the cross as being focused on reconciling people to God, but Paul here is broadening out our understanding of salvation to include *all things,* not only human beings. The God who created the world is the same God who redeems us and his whole creation through his Son, Jesus Christ. As God spoke the world into being, so Jesus is the Word who was there in the beginning (John 1.1), the one through whom and for whom and in whom all things were created (Col. 1.16).

In the English language there is no single word that brings creation and redemption together in Jesus, and westernized Christianity traditionally struggles to hold these two concepts together, though the Cherokee Indians have a beautiful phrase: *Oo-nay-thla-nah-hee Yo-way-jee*. This description of Jesus can only inadequately be translated as Creator-Son. Randy Woodley, speaking from his personal experience as a Keetoowah Cherokee, says that 'in this simple formula Jesus is acknowledged as both divine Creator and divine Son'.[7] There is a danger in our churches that we focus on the God of salvation and forget the God of creation. I wonder if this is something you recognize? Do you see that in your church or in your own thinking and awareness? Our task is to learn how to hold both concepts together and see how this might turn our attention outwards and change the way we live.

There will be opportunity to consider our response many times during this book, and we will look particularly at the place of human beings in relation to the wider creation when we reach Day Six in Chapter Six. For now, let us circle back to God as the creator of all that is, seen and unseen. Creation starts not with that which is created, but with the supreme God, who does not need to struggle and battle to bring forth life, but speaks the word and it is so. God is the one who begins by saying yes to life.

Let there be light

The opening verses of Genesis 1 speak of a dark empty nothingness. The Hebrew words used are *tĕhôm*, meaning the deep (relating linguistically to *tiamat*[8]), and *tōhû wa bōhû*, translated as 'formless and empty' (NIV) or 'formless and void' (NASB). Opinions vary over what precisely the author is describing, whether literal nothingness or a state of chaos, a shapeless mass of raw material, waiting to be formed and inhabited. However we view it, into this condition God speaks, 'Let there be light'. On this first day creation is spoken into existence.

To the Jewish mystics, God's word is not a once-for-all speech act which summoned life and then stopped. In a personal conversation with Rabbi Jonathan Wittenberg, he told me, 'God's voice in nature is ceaseless and enduring. God's speech is the invisible pulse which ceaselessly imparts vitality to all existence; it is the mute resonance, the silent song which is the essence of all consciousness.' Because of this, 'the refrain which opens each day of creation should not be translated as "God said", but rather "God says". Were that voice, that sacred energy to cease, nothing could survive.'

It is worth noting here the symmetry and order with which the author sets out his creation account. In the first three days, God separates and creates the spaces, the environments for created beings to inhabit, and then on Days Four to Six he creates the beings that will fill those spaces. So on Day One he separates light from darkness and creates night and day, which the sun, moon and stars of Day Four will fill. On Day Two he separates the waters of the sky from the water below, creating the sky and the seas, which the birds and sea creatures of Day Five will fill. And then on Day Three, God separates the land from the water, creating dry ground that the land animals of Day Six will inhabit, including human beings. Days One and Seven act like bookends, topping and tailing the account, with Day One functioning like a title and Day Seven being the conclusion as God rests from his creative work. It is beautifully done.

The Genesis 1 Seven-Day Creation Account

'formless . . . and empty' (v. 2)

Day 1	Day 4
Light; day and night	Sun, moon and stars
Day 2	Day 5
Waters above (sky); waters below	Birds and sea creatures
Day 3	Day 6
Land separated from sea	Land animals
Day 7 God's rest	

Alongside water, created on the second day, light is the lifeforce of the world. God separates the light from the darkness and declares that the light is good (v. 4). Light is a powerful symbol throughout the Bible, so let's take a few moments to explore this. Fundamentally light is used to symbolize God and God's presence. The psalmist proclaims 'The Lord is my light and my salvation' (Ps. 27.1) and when the plague of darkness covered the land of Egypt, 'yet all the Israelites had light in the places where they lived' because God was with them (Ex. 10.23). God's presence as light acts as guidance in the darkness to show his people the way and how to follow him. This is perhaps seen most literally when the Israelites are led through the wilderness as they flee captivity in Egypt under Pharaoh. As they embark on their perilous journey, God goes before them and with them, in a pillar of fire by night and a pillar of cloud by day (Ex. 13.21–22). In a more figurative sense, the psalmist famously declares of God's laws: 'Your word is a lamp for my feet, a light on my path' (Ps. 119.105), describing the way that God's word illuminates the way we should follow and provides guidance on how to live our lives.

Light as the presence of God is carried into the New Testament with John declaring 'God is light' (1 John 1.5) and James describing God as 'the Father of the heavenly lights who does not change like shifting shadows' (James 1.17). As the light shines in the darkness, revealing what is there, so nothing can stay hidden in the presence of God. Light thus comes to symbolize holiness and purity – we walk in the light as God is in the light (1 John 1.6). Therefore Paul calls us to 'live as children of light for the fruit of the light consists in all goodness, righteousness and truth' (Eph. 5.8–9).

Although Timothy can state that God 'lives in unapproachable light' (1 Tim. 6.16), the good news is that, through Jesus Christ, we *can* draw near to the God of light and know him because Jesus himself is light, and if we live in him, then we live in the Father. As Paul says, 'For God, who said, "Let light shine out of darkness", made his light shine in our hearts to give us the light of the knowledge of God's glory displayed in the face of Christ' (2 Cor. 4.6).

Jesus is the light and life of the world; 'the radiance of God's glory' (Heb. 1.3). He is the Word made flesh, and the same power of the word of God – that brings into existence what he speaks and that brings order out of the darkness of chaos – is in Jesus too. In John 9 when he declares that he is the light of the world (as he does also in 8.12), Jesus demonstrates the reality of that in the physical realm by healing a man born blind. He spits on the ground – mingling his very being with the dust of the earth – rubs the mud on the man's eyes and when the man washes the mud off in the Pool of Siloam . . . he can see!

Yet there is a note of judgment here, too. Light and dark cannot exist together and when the light of Jesus shines in the darkness, the darkness cannot overcome it (John 1.3). After Jesus heals the man born blind, the Pharisees object to the healing and throw the man out. Jesus goes to find and tell him, 'For judgment I have come into this world, so that the blind will see and those who see will become blind' (v. 39). The light is not good news for those who want to live in darkness!

Jesus came to bring light to everyone who will receive it, and this has both a present focus in the call to walk in the light, and a future focus as we look forward to a time when we will live fully in the glory of God with Jesus the Lamb as our lamp (Rev. 21.23). In that time, we will not need the light of a lamp or the light of the sun, for the Lord God will give us light (Rev. 22.5).

Lent is associated with both light and darkness. For those living in the northern hemisphere, it comes when the days are getting longer and the nights shorter; when the light is bringing life as new growth appears and the scent of hope is in the air. The promise of Easter and the resurrection is all around, leading us through the dark nights and reminding us that soon the light will come. This symbolism is used for the Maundy Thursday service at Chichester Cathedral in the UK. At the end of the service, the clergy and congregation process to the back of the church to a small garden area created to represent the Garden of Gethsemane, where Jesus prayed near his disciples as he was waiting to be arrested and crucified. The lights are turned off as we move to the garden, and as candles are lit, we symbolically take our place in the darkness among the disciples, waiting for Jesus through the night. For those living in the southern hemisphere of course it is the other way round, and the summer days are giving way to autumn as darkness increases and life begins to go underground. The focus on repentance – recognizing where we have walked in darkness and putting to death the old ways of life – fits in well with the darkening days, and we look forward to Easter breathing the Light of the World into our lives.

Whatever part of the globe you are in as you read this, pause for a moment to consider the illumination around you. Maybe it is evening and you are under an electric light. Maybe it is morning and the dawn is just breaking. Maybe the sun is glaring hotly as it always does, or maybe it is cloudy outside. Ask God, who speaks light and life into being, about the areas of light and life that he wants to speak into your being and into this world. What does he say to you?

As you contemplate, you might like to pray this Celtic prayer by priest and chaplain, Tess Ward:

Blessed be you Light of Life,
source of the sacred flame within each of us,
light which the darkness cannot put out.
I rise up with you this day.
I rest with you this night.[9]

Into the light

Moving from our biblical exploration to thinking about light today, let me tell you about an uncle of mine who is a lover of light. Uncle John was at the forefront of the silicon valley revolution. He was friends with Bill Hewlett and David Packard and remembers the evening, over a cup of 'hot tea', when his friend Bob Noyce told him he had decided to leave his current company and start a new one called Intel. He also recalls, some time later, Bob revealing he had come up with something called the integrated circuit that was 'going to change everything'.

Picking up the invention of an ageing uncle, John developed the first direct reading spectroscope: an instrument that splits light into its different wavelengths in order to identify the materials or elements that make up an object. It can be used for all sorts of things, from finding out the physical properties of objects in space to determining the quality of gemstones. John's incredible life has included a childhood in a prison of war camp in China; a world-famous trek across the Alps with an elephant called Hannibal as a student, and his life in silicon valley. He bases his autobiography, *Persistence of Light*, on the seven colours of the rainbow,[10] which provide a framework for his life story; each unique and beautiful, and, all together, making white light complete:

- Red is the color – red for China, red for violence, red for the heart of a family torn apart.
- Orange is the color – the blending of red and yellow. These years brought the red of my experiences in the Japanese camp in China to the English mix of high school and army, a preparation for the joyful yellow of college.
- Yellow is the color – for the joy of being a student in such an unforgettable place and for the promise of new adventure.
- Green is the colour – for the rich greens of nature, for the deep woods of the Alps. It fits happily and cheerfully between blue and yellow. How much of each determines the varied shades of green, reflecting the many moods of our elephant expedition: the emotional highs and lows, the dangers, and the bliss.

- Blue is the color – for the vast horizons of possibility and opportunity. It is often used to convey creativity and intelligence.
- Indigo is the color between primary blue and violet. It is also the name of my step great-granddaughter, symbolizing for me the love of our growing and extended family.
- Violet is the color at the far end of the visible spectrum. It is what we might see just after the sun sets.

Light is the foundation of life: the essential building-block, created on the First Day, emanating from the Word and the Spirit of God. It is so foundational that we can live in it and yet miss its beauty and wonder. We are in fact completely dependent on light. We need it to grow our food and give us Vitamin D to keep our bones healthy and strong. We need it to maintain our body clock: to tell us when to be awake and when to sleep. Some people are affected by Seasonal Affective Disorder (SAD) that leads to depression in the winter months from lack of warmth and sunlight (those of you reading this in hotter climes can hardly imagine what that must be like!) We need light to send images of things around us to our brains so our eyes can see and, without light, there is no colour. While we will not see God face-to-face until Jesus comes again, light enables us to perceive the world he has created that shows us something of his beauty. Moreover, we catch a glimpse of how God views the world when we are able to step back and marvel at his creation. As we will see again in Chapter Three, the refrain throughout Genesis 1, that 'God *saw* that it was good [emphasis added]', demonstrates that he took time to look at and appreciate the inherent worth of the work of his hands in creation. In providing us with light so we can see, he demonstrates his intention for us to share this enjoyment with him – to perceive that the world is good and to declare it to be so.

Moreover, we are not the only ones who respond to light. Deciduous trees seem able to 'count' the number of warm days until there have been enough for them to be certain it is safe to start growing new leaves. But the warm days are not enough and trees need light too. Beeches, for example, begin their growth once it has been light for at least thirteen hours a day.[11]

We see something similar in birds. They respond to light in their breeding habits, needing the days to have so many hours of sunlight before hormones are kick-started and the breeding process begins. It used to be thought that this was not relevant in the tropics where day-length hardly varies, but scientists have found otherwise. The spotted antbird (*Hylophylax naevioides*) is a small tropical bird that lives in Central America. There, daylight varies by just one hour through the year. At twelve hours of daylight the males display no signs of hormonal activity, but by thirteen hours they are fully sexually active. Closer research seems to show that just seventeen extra minutes of daylight after twelve hours prompts the beginning of changes in the male.[12]

In fact, when it comes to light we need some humility because there are other species far more receptive to it than are we. As humans we do not perceive all the light that is seen by other creatures. Animals use ultraviolet light which reveals whole areas of living and being that are by nature inaccessible to us. Kestrels, for example, use ultraviolet light to find their prey. Voles mark their runways through fields with urine and faeces which are visible in ultraviolet light. Wild kestrels brought into captivity were able to detect vole scent marks in ultraviolet light but not in visible light, which suggests they can see through the grass to the ultraviolet trails beneath it though these are not apparent to the human eye.[13]

One amazing use of light in the natural world is bioluminescence. I remember as a child staying in a beach hut on the west coast of Malaysia with my parents while they were on a speaking tour there. One evening we walked onto the beach in the dark and the shoreline shimmered with phosphorous, changing patterns as the waves lapped along the shore. I paddled out and as I lifted my hands and splashed the water into the air, brilliant drop of iridescence fell around me. On another trip to Malaysia I was taken out at night to a village along a river in a very rural area with no electricity. We rowed out into the blackness and down the river, and gradually the bushes around us came alive with the twinkling lights of fireflies. We kept silence watching the magnificent display.

In the seas, bioluminescence is seen not just on the surface of the water (where it is caused by a mass of bioluminescent microorganisms), but in all sorts of other living creatures underneath too. The anglerfish dangles a lure in front of her that flashes and attracts prey, while the vampire squid squirts a bioluminescent liquid rather than the black ink of most squid. Deep underwater, where light with longer wavelengths cannot reach, many deep sea creatures are red in colour, effectively making them invisible. However, some animals, such as the dragonfish, have developed the ability both to make and to see red light. They then emit that and use it to spot their prey.[14]

Lighting up the darkness

> 'God saw that the light was good, and he separated the light from the darkness. God called the light "day", and the darkness he called "night".'

One of the many fascinating things that NASA has produced in recent years is its imagery of the earth at night. In partnership with the National Oceanic and Atmospheric Administration (NOAA), it used the satellite imagery of 312 orbits around the world to put together a view of what the world looked like. What emerged was a stunning view of 'the black marble'.[15]

This satellite imagery is used for all sorts of weather-related and military research, but also for things as varied as calculating the CO_2 emissions from wildfires and gas flares; monitoring the activities of commercial fishing fleets who use lights at night, and assessing how urban growth has broken up animal habitats. What the NASA scientists discovered is that, 'unlike humans, the Earth never sleeps'. Even away from human habitation 'light still shines. Wildfires and volcanoes rage. Oil and gas wells burn like candles. Auroras dance across the polar skies. Moonlight and starlight reflect off the water, snow, clouds, and deserts. Even the air and ocean sometimes glow' . . . 'the night is nowhere near as dark as most of us think. In fact, Earth is never really dark; it twinkles with lights from humans and nature.'[16]

It is these 'lights from humans' that have really stood out as, for the first time, we have been able to see cities across the world at night and gain some appreciation of just how much light is being emitted. As

Chris Elvidge, a NOAA scientist who has studied this for 20 years, says, 'Nothing tells us more about the spread of humans across the Earth than city lights'. These images have shown the increasing urbanization of the human population and, through the ability to measure the emissions from lights, has enabled a growing understanding of humanity's footprint on the earth.[17]

We are now aware of the huge amount of light being produced and the vast inequality between those who have it and those who do not. South Korea, for example, is brightly lit while North Korea is in almost total darkness. Also immediately obvious is the colossal amount of electricity being used to power all those lights. As the narrator of NASA's 'Earth at Night' video says, 'The night is electric.'[18]

We are inside the light!

Electricity is hugely important in our world today due to its role in providing us with light, and also more generally because of the significant effect its presence or absence has on the quality of people's lives. Phul Kumari is 32 years old and lives with her husband, three daughters and son in Kalanga, a village in Nepal. In the past, the village had no electricity and this had a massively detrimental impact on the life of Phul and her family. She had to get up at 4am to grind their maize, wheat and corn by hand which took a long time. For light, they relied on kerosene. Phul had to walk a long way to buy it, travelling up to four hours on foot, and it was so expensive that sometimes she found she could not afford it. To help make ends meet, her children also got up early in the morning to work. At night, they had only the dim, smoky light of the kerosene lamp to study by which often meant they didn't finish their homework; they were getting behind with their studies and their education was suffering.

The good news is that the village now has a micro-hydropower plant, thanks to an organization called United Mission to Nepal (UMN), which generates electricity to provide light to the community. This has transformed everything! 'Before having hydroelectricity, our community felt like our lives were on a hard trajectory but after having electricity, we feel very happy because we are inside the light.' With electric light at home, Phul's children are now getting all their homework done, their rate of school attendance is increasing and their education is improving. Phul is a member of the school management committee and regularly hears from teachers about her children's good progress. 'I want my children to have a different life. They might be engineers or doctors. I hope my children will do good in the community.' Phul can now also bring the family's grain to the newly opened mill – which uses water run-off from the plant – where the owner will grind it in just an hour. Now that she has the time – and the inspiration from other businesses opening in the village – Phul is starting her own poultry and vegetable farm. The micro-hydro plant provides light for the farm and heating to keep the chickens warm. 'Now I can engage in business. I support my children so they can go to school.'[19]

Affordable and clean energy has been identified as one of the Sustainable Development Goals (number 7). As António Guterres, Secretary-General of the United Nations, has said, 'Energy is the golden thread that connects all the sustainable development goals.'[20] Traditional fuels like kerosene and candles are dangerous as they can cause accidents and create fire hazards. In Nigeria, almost a third of hospital burn patients have been injured by fuel exploding while using kerosene lamps. Children can also suffer from poisoning after drinking kerosene by mistake as it is sold in bottles that look like water bottles, and it is a horrible air polluter.[21] Around three billion people still cook by burning biomass like wood and charcoal, and the indoor air pollution from that contributes to around four million premature deaths each year, primarily among women and children, as well as being another serious burns hazard.[22] In fact, research has shown that air pollution kills more people each year than malaria and HIV/AIDS combined.[23]

People need clean, affordable energy to get out of poverty. Renewable energy like solar panels, which is often the cheapest source of light, means cleaner air, new jobs, schools and universities with lighting and computer access. Farmers can irrigate fields with electric pumps and increase their yields. Children and students can study in the evening. Electrification in clinics and hospitals lengthens treatment hours, provides access to equipment and improves surgery. Women can give birth safely at night and vaccines can be kept in refrigerators. Businesses can run smoothly and increase their profits.[24]

Ali is 35 years old and lives in Kinangali village in the Dodoma region of Tanzania with his wife and two children, aged 12 and 4. They moved to the village four years ago. When he arrived, Ali identified a need in the village: local people kept cattle, but had to travel a long distance to Manyoni for treatment for their animals if they became sick. He opened a kiosk selling medical supplies for cattle, such as dipping solution and antibiotics. Ali had been thinking of moving again until he heard about the launch of something called Pamoja, the name given to the community savings groups that Tearfund has initiated in Tanzania.[25] He got involved in a group with nineteen other people and decided to stay. He took out a loan from the group, and with 75,000 TZS (33 USD) bought a solar panel. The plans the government has for the national grid will not reach Ali's village for many years. In the past, Ali used a torch with batteries for light and had to close his kiosk at 6pm when it got dark. Now, with the solar panel, he can stay open in the evenings; he can also charge his mobile phone, which he uses to buy goods for his business. Ali used to earn 5,000 TZS (2 USD) a day; now he makes 10–20,000 TZS (4–8 USD). With the extra money, he has bought a plot of land, plans to buy a house and would like to expand his business. With a more powerful solar panel, he would generate enough electricity to buy a fridge to sell cold drinks or to charge other people's mobiles. The Pamoja group enables the local community to save money together and share in their successes: 'I will continue to be in this Pamoja group so I can prosper and tell others to join the group because I know for sure we will start at the bottom and go to the top – step by step we will succeed in our lives.'[26]

Access to light and electricity is hugely important, and the good news is we have seen giant leaps forward over recent decades, with the latest figures putting the global electrification rate at 89 per cent. More than 920 million people have gained access since 2010. However, there are still about 840 million people without electricity and, unsurprisingly, progress is uneven. In some instances, even where people are connected, cost can still be a problem, meaning they continue to use harmful fuel sources like kerosene or charcoal which are more affordable.[27] Electrification efforts have been particularly successful in Central and Southern Asia; Latin America and the Caribbean; and Eastern and Southeastern Asia, which are all between 91–98 per cent of access. But, sub-Saharan Africa still faces immense energy challenges and, across that region, more than half the population – 56 per cent – does not have access to electricity. It is projected that 650 million people are still likely to be without electricity in 2030, and nine out of ten of those people will be in sub-Saharan Africa.[28]

While access to light, electricity and energy in general is crucial, it is also important to consider the source of that energy. According to the latest figures (at the time of writing), 65.1 per cent of electricity around the world comes from fossil fuels; 16.6 per cent from hydroelectric plants; 10.4 per cent from nuclear plants; 5.6 per cent from geo-thermal, solar, wind and tide; and 2.3 per cent from biofuels and waste.[29] This is important for a variety of reasons. Tearfund's report with the ODI (Overseas Development Institute), *Pioneering Power: Transforming energy through off-grid renewable electricity in Africa and Asia*, shows that solar and micro-hydropower are often cheaper, faster, more reliable, safer and cleaner than extending a centralized grid or using diesel or kerosene. And renewable technologies could employ 1.8 million people in sub-Saharan Africa alone. Off-grid renewable energy offers clean development and it provides a way to 'leap-frog' over traditional fossil-fuel based power lines to a much more sustainable option, that enables people to be lifted out of poverty without expense to the environment.[30]

This is incredibly important because of the role fossil fuels have played in causing the climate emergency we are facing. Climate change is going to be a theme we will encounter at various points in this book because of the terrible impact it is having on God's creation, both people and planet. But as we look at the theme of light in this chapter, we cannot help but be aware of how the provision of light to so many billions of people, which has brought such incredible benefits, has also caused incredible damage.

In 2018, the IPCC (Intergovernmental Panel on Climate Change) produced a Special Report examining the impact on the earth of 1.5°C of warming above pre-industrial levels. It showed that above 1.5 degrees we are facing increasingly extreme weather events, sea level rise, species extinction and melting glaciers. From a human perspective, health, livelihoods, food provision, water supply, human security and economic growth are all seriously at risk.

The report assessed the feasibility of staying under 1.5° and what we need to do to achieve that, including the costs involved. Its main messages were that limiting warming to 1.5° is attainable, and that every fraction of a degree will make a difference to the effects on the planet. According to the report, the consequences of a 2° warmer world are far greater than that of 1.5° – though the impact of the latter is still severe. The report also noted that the world has already reached 1° of warming, and that this warming is already affecting us. Currently we are on a path towards a 3° warmer world.

The report was clear that we will only change the current trajectory we are on and keep our world to within 1.5°C of warming if we take urgent action. This means moving on from the incremental changes that have been happening to transformational change. The report gives us ten years to alter course, and says that the speed at which we do or do not cut our emissions between now and 2030 will make or break our prospects of keeping temperature rise to 1.5°. The world is way off track from reaching net zero even as late as 2050, and it is increasingly clear that even that goal is too late for millions of people.[31] If we do not reduce our carbon emissions by 45 per cent in the next ten years,[32] it is likely we will see 100 million people being pushed back into poverty, global crop yield losses of as much as five per cent over the next ten years, and the disappearance of all our coral reefs.[33]

Former Archbishop of Polynesia, Winston Halapua, expresses things well when he says, 'The threat to survival from climate change is unparalleled and ongoing'. He sees first-hand the impact that it is having on the low-lying islands of Oceania and says that, because of climate change, 'Powerless people [are] being violently abused in the environment of their own island homes'. He calls on 'other parts of the world to acknowledge that what happens to islands in the Pacific Ocean has far reaching significance'.[34] Moving swiftly towards net zero emissions is crucial in responding to his call.

A friend of mine, David Nussbaum, used to be CEO for World Wildlife Fund UK and is now CEO of The Elders. He remembers a conversation he had some years back with the then UK Energy Minister of what was called DECC (the Department of Energy and Climate Change). Towards the end of the conversation David was asked if he believed in God and he replied yes, and that he sometimes wondered what God thought about climate change and our use of fossil fuels which are contributing to it. He goes on:

> I suggested that perhaps God saw it this way: he arranged for the earth to have some coal and oil easily accessible, so that we could get the industrial revolution going – with all the benefits that brought – but ensured that most of the rest was buried deep down where it was difficult, dangerous and expensive to get at it. But, he'd also ensured that the sun and moon were staring us in the face with sources of energy, and the wind was blowing all around us. So maybe God was puzzled as to why, especially now we knew about climate change, we hadn't taken the hint to use the easily accessible and safer sources of energy rather than the inaccessible and dangerous ones![35]

Electricity is one of the largest sources of CO_2 emissions, and if we are to provide light to the world – as well as heat, cool air and cooking fuel – we will have to do that in ways that use renewable sources and neither pollute the environment nor pump greenhouse gases into the atmosphere.

Words, words and more words

You may well now be feeling overwhelmed because the changes needed in this situation are huge and way beyond anything any of us reading this can do on our own. I confess to feeling like that often! But we have seen in this chapter that God is the creator of this world and that Jesus died to reconcile to God all things, both people and the whole creation. Therefore, as part of our worship, we want to respond to the needs of people living in poverty and to the needs of the wider creation, which is in desperate trouble because of the climate emergency. If we are to move away from fossil fuels in the way we need to, and enable people to keep the lights on, *all* of us are required to act. As Thabo Makgoba, Archbishop of Southern Africa, says, 'Words, words and more words will not reverse environmental degradation of carbon emissions, but our actions together can. So act now for climate justice. Change begins with us!'[36] And my prayer is that all of us reading this chapter, in whatever country we are and whatever denomination or church network we are a part of, will be inspired to take action, both individually and in our churches.

The first thing we can do is **learn**: be informed and gain knowledge. There are excellent websites, magazines, newfeeds and organizations that will help you sort fact from fiction, increase your understanding, and help you act in the most effective ways. Some of these you will find at <www.spckpublishing.co.uk/saying-yes-resources>, but do your own looking around to find the things that work best for you.

Second, we can **take action in our lives** to live in ways that reduce our own carbon emissions. Readers of this book are from many different countries and so it is hard to generalize. Key actions will look different depending on where you live and your individual circumstances. In general, on a global level, the big thing is that we need to change our eating habits so we consume a predominantly vegetable and grain-based diet. We may wish to include a small amount of meat and dairy (recognizing that there is a big difference between animals raised in an intensive feedlot system and animals reared on a more natural diet). Also important is to choose less polluting means of travel, such as public transport, car shares or to invest in an electric car (where the latter is an option), as well as significantly reducing the amount we fly (if not cutting it out altogether). We can choose a domestic energy supplier that uses renewable energy; perhaps install our own solar panels; and plant trees whenever we can. In thinking about our electricity use specifically, we must reduce the amount of energy we consume by favouring energy-efficient lights and appliances and making sure we turn things off when we are not using them.

We can also take action in our churches by bringing the situation into our prayers and sermons, and helping people understand why taking action on climate breakdown is a Christian thing to do. In practical terms, Stratford-Upon-Avon Methodist Church in the UK used the opportunity of needing to refurbish the church to significantly reduce their gas consumption by installing solar panels and a ground source heat pump. They also started using rainwater to flush the toilets and overall reduced the carbon footprint of the church premises from 40 tonnes to 16 tonnes of CO_2 annually. Though seeing an increase in the number of community groups using the building, they have still managed to reduce their environmental impact.

Meanwhile in Kenya, Benjamin Kyalo, who is a member of the Diocese of Machakos, became concerned about the environmental degradation in his area, including the effect of deforestation on the land and its contribution to climate change. He decided to do something and began to plant tree seedlings. He is now cultivating four thousand seedlings at his house, selling them at a small cost to neighbours, local parishes and the diocese. In recognition of his work to promote the protection of the local environment, he was selected as the youth representative of the Anglican Church of Kenya to the Anglican Consultative Council at their 2019 meeting in Hong Kong.

Another initiative seeking to engage Christians in planting trees is being run by Bishop Ellinah Wamukoya, the first female Anglican bishop in Africa, who is passionate about caring for the environment, in the Diocese of Eswatini (Swaziland). She has adopted the practice of marking important spiritual events (such as baptisms, confirmations, weddings and funerals) by planting trees, and young people in the diocese are taking the lead. Alongside this, the diocese has started a project of planting wattle trees in areas prone to soil erosion, and it plans to add indigenous trees to these plantations in future to combat the spread of other invasive species. In recognition of their efforts, the diocese has won an award for being the most environmentally active faith-based organization in the country twice, in 2016 and 2018.

Third, we can **use our voices** to push governments and businesses to make the large-scale transformatory changes that are needed. At the Paris climate change talks in 2015, 196 countries signed an agreement to cut greenhouse gas emissions and phase out fossil fuels in order to keep temperature rise to 'well below two degrees'. We need to hold our governments accountable to that and keep asking them what progress is being made and pushing them to act.

Finally, for those of us in a situation to do so – as individuals and in our churches – we can **give** our money to support organizations working both to help communities adapt to climate breakdown and mitigate further changes. For those of us reading this in the UK, CAFOD, Christian Aid and Tearfund all work in these areas and our financial support will really help make a difference on the ground and at a global level. But, wherever in the world you are based, your church may well be in a financial situation to support organizations who are helping vulnerable communities learn new climate-resilient farming methods or invest in renewable energy (for example), and I believe it is part of our Christian calling to use our money to fund this work.

Together there is much that we can – and indeed must – do and you will find lots of further, practical information on the resources page online at <www.spckpublishing.co.uk/saying-yes-resources>, so do take a look there. Leading atmospheric scientist Canadian Katharine Hayhoe says, 'For Christians, doing something about climate change is about living out our faith – caring for those who need help, our neighbors here at home or on the other side of the world, and taking responsibility for this planet that God created and entrusted to us. My faith tells me that God does want people to understand climate change and do something about it. And that is a very freeing thought: I don't have to change the world all by myself, I just need to partner in the work God wants us to do'.[37]

God saw that the light was good

And so we start our Lenten journey with the theme of light, and as we respond to what we have looked at here – whether on our own or with a Lent group – let us begin by stepping into God's light and asking him to open our eyes to the many dark places, both in our own lives and in the world. We need God to illuminate the darkness. Archbishop Desmond Tutu has said, 'Good is stronger than evil; love is stronger than hate; light is stronger than darkness; life is stronger than death. Victory is ours, through him who loves us'. In that knowledge, may we be carriers of light.

For discussion

1 In the Introduction and in the first part of Chapter One, we looked at the context within which Genesis 1 most likely found its final form. What did you think of that? How does this affect your understanding of this first biblical account of creation?
2 Earlier in this chapter, Wangari Maathai remarked, 'One is left to wonder whether conceiving of God as the origination of all that is would make people of faith recognize that they have a responsibility to be the custodians of God's creation and, in the process, their own survival'. Take time to reflect on this. What place does God as creator have in your faith and in that of your church? How might it make a difference if that understanding was more strongly emphasized?
3 Where does light feature in your own life? Can you share any particular experiences that have given you a new appreciation of light?
4 For this chapter we are privileged to have an interview with Christiana Figueres from Costa Rica who, for six years, was Executive Secretary of the UN Framework Convention on Climate Change (UNFCC) and oversaw the 2015 Paris Agreement. Watch her interview at <www.spckpublishing.co.uk/saying-yes-resources>.
5 All of us need to make changes if the climate crisis is to be averted. Go back and look at the action points towards the end of this chapter and, if you have access to the internet, look at the online resources. What changes will you make? When will you make them?

A prayer on light from the United States of America

To the one who surges into the fissures that cannot be accessed,
To the one who floods into the crevices that cannot be reached,
To the one whose presence exposes, clarifies, restores, and cultivates,
We bask. We thrive. We harbor no secrets. We hold onto no heaviness we fear makes us unloved.
You are the first ingredient for life.
The universe, not to mention our hearts, would wither away without you, Light of the world.
Amen.

Morgan Lee is a journalist and community organizer who lives in Chicago. She loves baseball, biking and learning new languages.

2

Let the waters be separated (Genesis 1.6–8)

[6]And God said, "Let there be a vault between the waters to separate water from water." [7]So God made the vault and separated the water under the vault from the water above it. And it was so. [8]God called the vault "sky." And there was evening, and there was morning—the second day.

My childhood memories are full of water. I was brought up at the stunningly beautiful All Nations Christian College in England, set in acres of Hertfordshire fields and woodlands. Behind the college, at the bottom of a field, winds the tiny River Ash, which we called Lilo Creek. We had lots of fun times as children splashing about in the water on our lilos and in a little yellow dinghy that my mum romantically called Yellow Dawn. Nearby was a ford where the cattle crossed – another fun place to paddle and play.

The area has lots of former gravel pits, by then turned into lakes, and my parents bought canoes. Many weekends were spent out on the lake. I could canoe before I could swim (probably not the safest thing looking back, especially as we never wore lifejackets!) I can remember adventures with my brother, canoeing through reedbeds into hidden areas and once discovering an abandoned rusty boat on its side, which we clambered onto and explored.

That childhood love of water has stayed with me into adulthood and near (or in) fresh water is one of my favourite places to be. My sister's house backs right onto a lake and we spend many hours there as an extended family. One of our best times recently was a scorching hot summer's day when we were all there, including my brother and his crew. We went out onto the lake in wooden rowing boats and spent the day jumping into the water. My sister and I swam together for a long time through the clear water, talking and sharing deeply (and putting out of our minds the 1.5 meter-long pike lurking beneath us!)

In this chapter we are celebrating water (focusing particularly on fresh water as we will look at the seas and oceans in Chapter Five). We will consider water in the Bible, see the role it plays for the people of God both physically and metaphorically, and then look further at the refrain, 'And God said . . .', considering what we can learn about why God made the heavens and the earth. We will then think about water today, be inspired to notice it and not take it for granted, and become more aware of the problems it faces and the part we can play in looking after it.

Separating the waters

In Genesis 1.6–8, God speaks and calls into being 'a vault between the waters to separate the water from water'. This vault, also translated as 'firmament' or 'expanse', is the creation of the sky and sees God bringing control and order to the swirling, chaotic mass of deep waters mentioned in verse 2.

The verb underlying the Hebrew for 'vault' (*rāqîʿa*) means to beat or stamp, as in beating out a sheet of metal.[1] It is the same word used by Elihu when he asks Job, 'can you join [Yahweh] in spreading out the skies, hard as a mirror of cast bronze?' (Job 37.18) and reflects an ancient view of the world, different from our own understanding. The process of separating out, outlined in the previous chapter, continues as the watery chaos that overflows the earth is tamed and divided to create space between the waters of the ground and the water that comes from the sky. Now there is space to inhabit and air to breathe.

Alongside light, water is an essential element for life.

Streams of living water

Water is a powerful symbol throughout the Bible. This reflects the fact that the Bible was written by people who were intensely aware of how precious water was and of the ever-present threat of it running out –

far more so than some of us reading this. Thus they lived with a very close day-to-day connection with water.

This awareness in turn made them more appreciative of water as a gift from God, his provision for his creation. In fact, Jewish thinking says that God specifically put his people in a land with no major rivers precisely to help them remember that God was the ultimate provider of water and of all their needs. In this way, Rabbi Yonatan Neril says, 'The Biblical experiences with water in the desert can be understood as a spiritual training to cultivate appreciation for God's goodness'.[2]

Water is a key theme in Psalm 65, which is a wonderful song of praise to the God who forgives and answers prayer and fills his people with good things. The whole earth praises God, lost in awe at the amazing things he does, and the earth is filled with songs of joy. The joy comes because of God's bounteous provision, and central to that is the role of rain:

> You care for the land and water it,
>   you enrich it abundantly.
> The streams of God are filled with water
>   to provide the people with corn,
>   for so you have ordained it.
> You drench its furrows and level its ridges,
>   you soften it with showers and bless its crops,
> You crown the year with your bounty,
>   and your carts overflow with abundance.
> The grasslands of the wilderness overflow,
>   the hills are clothed with gladness,
> The meadows are covered with flocks
>   and the valleys are mantled with corn,
>   they shout for joy and sing. (vv. 9–13)

Have we lost that sense of rain as a provision from God? It is dry in the south east of England where I live and we have had very little rain recently. In fact the Environment Agency is warning that England could run out of water within 25 years because of rising population and climate breakdown – something that readers from other places might find inconceivable given the stereotype of England as a rain-soaked land of black umbrellas![3] And of course, many other parts of the world are facing much more extreme conditions, which we will reflect on later in this chapter. I found myself musing on the lack of rain as I walked around my area recently, confronting the reality of my absolute lack of control over whether or not it rains (though of course I can control what happens to the water when we do have showers, and whether I use it wisely or not).

Yet the Scriptures are clear that water is part of God's provision for his creation, including his people. In Genesis 21, God compassionately provides for Hagar in her situation of deepest need. In this story of water injustice, we see Hagar thrown out into the desert by Abraham and Sarah. The one skin of water she was given has run out, and she now sits down some distance away from her son, Ishmael, to wait for him to die. She is in tears, alone and abandoned. But God comes to her, opens her eyes and shows her a well of water. God is with mother and son and Ishmael grows up to become the founder of a nation of his own (Gen. 21.8–20).

Jesus tells us that God can send rain on whom he likes, both righteous and unrighteous (Matthew 5.45) – but he can also withhold rain too. In the time of King Ahab, in judgment against him, Elijah announces that 'there will be neither dew nor rain in the next few years except at my word' (1 Kings 17.1), and so the rain stops and a terrible drought occurs. This sets the scene for the mighty showdown with the prophets of Baal on Mount Carmel. Elijah's use of water, pouring large jar-fulls over the offering, must have felt outrageously wasteful when the land was suffering such a famine because of the drought, but he is confident in God's promise that he would send rain again. And sure enough, as a sign of Elijah's victory over the false prophets – and more importantly of God's victory over Baal – the skies grew black, the wind rose and a heavy rainstorm broke (1 Kings 18.45).

The composer Mendelssohn's depiction of this scene in his oratorio *Elijah* is tremendous, with Elijah sending out the servant to look towards the sea, searching for signs of rain. The tension builds up as, time after time he returns, reporting in his high treble voice, 'No, there is nothing', until finally, on his seventh return, he sings tremulously, 'Behold, a little cloud ariseth now, out of the water . . . It is like a man's hand'. And, as the orchestra picks up pace, you know the rains are coming!

Water has a shadow side to it. It can bring life but it can also bring death and suffering, whether through engulfing people (as in the flood of Genesis 7 or the Red Sea flowing back over the Egyptian army in Exodus 14); or through the earth drying up and provoking famine, due to lack of rain or a tactic of warfare (2 Kings 3.25); or through being unclean and undrinkable. The bitter water of Marah was changed into drinkable water by Moses throwing a piece of wood into it as God showed him to do (Exodus 15.22–26).

This shadow side is used in a metaphorical way too and the Psalms speak vividly of being in 'deep waters' while 'the torrents of destruction overwhelmed me' (Psalm 18.16 and 4). Whatever situation the writer of Psalm 69 is in, he describes it as feeling like 'the waters have come up to my neck', and he goes on to say:

> I sink in the miry depths,
> where there is no foothold.
> I have come into the deep waters,
> the floods engulf me.
> I am worn out calling for help,
> my throat is parched (vv. 2–3).

I have no doubt that all of us reading this can think back to times when those words have been our own, and maybe even now they have resonance for some of us in situations we are currently facing.

I recollect sitting down to pray in the midst of a dark and painful time. I closed my eyes and immediately and vividly saw a picture of myself sinking down into deep waters, way over my head. It matched exactly how I was feeling, as my circumstances seemed totally overwhelming and it appeared I was powerless to do anything to change them. I felt like I was drowning and would never resurface.

But, while recognizing this shadow side, the Bible affirms that God is in control of the waters and we will be safe. When the writer of Psalm 69 cries out to God, 'Do not let the floodwaters engulf me or the depths swallow me up', his experience is of rescue and salvation and he can declare, 'The Lord hears the needy and does not despise his captive people' (vv. 15 and 33).

Come to the Living Water

In the Gospels we see that Jesus is Lord over the wind and the waves, which obey him when he and the disciples are caught out on Lake Galilee in a storm (Mark 4.35–41). Of course they obey him: he is the one who created them, whose Spirit was brooding over the waters of chaos! He is the Lord of all creation. And even when we feel like we too are drowning in the depths, God is with us in his love. Father, Son and Holy Spirit are weaving together and weaving our struggles into the life at the heart of the Godhead, turning our grappling around to bring hope.

Returning to the picture of me sinking down, as I sat with God in prayer, he broadened the image out a step or two and I realized that my arm was held up and coming out of the water and there, standing on the bank, was Jesus, reaching down to hold my hand. I was still in the water – he hadn't pulled me out – but I knew he was stopping me sinking down any further and clasping me securely.

Jesus is at the centre when we consider how the theme of water flows through the Bible. We first encounter him and water in the story of his baptism by John in the River Jordan (Matt. 3.13–17; Mark 1.9–11; Luke 3.21–22; John 1.29–34). The Jordan is of course key to the story of the people of God: it is the river the Israelites crossed to leave behind the Exodus and enter into the promises of God to be his people and to have their own land. As with the Red Sea when they left Egypt, so here too God stops the water so all the Israelites could walk across on dry ground (Joshua 3, 4). The River Jordan therefore is the symbol of freedom: the sign that God's promises are being fulfilled.[4] And it is in this river that Jesus is baptized: *Yeshua*, Joshua, the anointed chosen one, the Messiah.

Through his baptism, Jesus is identified as God's beloved son and the one for whom John has been preparing the way. But Jesus will baptize with something far more powerful than water: the Holy Spirit. It is in

Jesus that the hopes of the Old Testament find their fulfillment; in Jesus the future promised by the Hebrew prophets comes together – a new creation characterized by God's full presence and the restoration of all things to his shalom. In the glimpses we are given of this transformed creation (for example in Isaiah 11 and Revelation 21 and 22), we see a picture of reconciliation with people being reconciled to God, to one another and with the natural world. We will explore this further in Chapter 4, but for now, as Jesus is baptized in the river Jordan, we see him confirmed as the one who brings that freedom and peace and restoration of relationships.

In our own baptism, we publicly step into that identity in Jesus as people of the new creation. Paul affirms, 'If anyone is in the Messiah, there is a new creation! Old things have gone, and look – everything has become new!' (2 Cor. 5.17, using Tom Wright's translation in *The New Testament for Everyone*). We become Jesus-people: signed up to his agenda of the Kingdom of God, committed to living out the values of justice, peace and righteousness in our everyday lives.

Baptism of course hinges on the literal and metaphorical role of water in cleansing and purification. Ritual washing was and is a key part of the faith of the Jewish people: priests had to bathe themselves before putting on the sacred garments on the Day of Atonement and before dressing in their ordinary clothes again. Aaron and his sons had to be washed before they were consecrated. Anyone who touched something or someone considered ritually unclean had to wash themselves in order to become clean.

Hundreds of years before Jesus' baptism in the Jordan, Naaman, the commander of the army of the king of Aram, was told by Elisha to wash seven times in the Jordan in order to have his leprosy healed (2 Kings 5). Of course, this was not because there was some magical property in the water that could bring healing, or because it was thought that leprosy could simply be washed away. It was not even because there was anything particularly mighty about the river Jordan itself: as Naaman says, 'Are not Abana and Pharpar, the rivers of Damascus, better than all the waters of Israel? Couldn't I wash in them and be cleansed?' No, this was about Naaman submitting to 'not just the tribal god of Israel, but the God of all the earth, the healing, liberating God, the God of creation and covenant, of Exodus, wilderness and Jordan, the God of steadfast redeeming love'.[5]

Baptism, then, is about going down into the waters in order to be purified from our wrong-doings; from all the things in our lives that go against the way God has created us to live – in shalom with his whole creation, both human and wider – and prevent us entering into the triune presence of God. It involves too that shadow side of water we saw earlier, as it is a type of drowning. Our old self is made dead, so that our new self might be made alive, resurrected with Christ.

I'm writing this on a Sunday afternoon, having been to my local cathedral service this morning. The Old Testament reading for today sums up wonderfully God's promise of what happens to us when we are baptized and brought into new life:

> I will sprinkle clean water on you, and you will be clean; I will cleanse you from all your impurities and from all your idols. I will give you a new heart and put a new spirit in you: I will remove from you your heart of stone and give you a heart of flesh. And I will put my Spirit in you and move you to follow my decrees and be careful to keep my laws. (Ezekiel 36.25–27)

It is through Jesus that the words of Ezekiel are brought to life in our lives as we submit to him and go under the waters of baptism. As Jesus tells the Samaritan woman he meets at the well – she who has had to collect her water in the heat of the day because she has been rejected by her village – he is the one who gives living water: 'whoever drinks the water I give them will never thirst. Indeed, the water I give them will become in them a spring of water welling up to eternal life' (John 4.14). The water that flows from Jesus continues into the vision that John has of the renewed heaven and earth: flowing from the throne of God and of the Lamb is 'the river of the water of life, as clear as crystal', and it pours down the great street of the holy city (Rev. 22.1–2).

Water is life and blessing, and the living water that Jesus offers brings blessing and refreshment right to the core of our being. Some of us reading this are in the privileged position of having rarely if ever experienced what it is like to be truly thirsty. Psalm 63 gives us an insight through the psalmist's longing after God: 'You God are my God, earnestly I seek you; I thirst for you, my whole being longs for you, in a dry and parched land where there is no water' (v. 1). Can we imagine what it is like to thirst so desperately, and can we ask God to create in us that same longing for his living water?

Some years ago, the church I am part of experienced an amazing time of heightened experience of God's power, with lives touched deeply and miraculous healings occurring more frequently than normal. One afternoon some friends were at our house for coffee, and as we were chatting, it started to rain. I felt an urge to go and stand outside and, as it rained harder and harder, puddles developed along the path in my back garden. I felt another urge to lie down . . . so I did, feeling somewhat embarrassed ('excuse me, just ignore me, but I feel I need to lie in a puddle for a while') and I lay there getting absolutely saturated, asking God to soak me with his presence and drench me with his love. I was saying yes to his life!

In Isaiah, God calls out an invitation to all of us who are thirsty: 'come to the waters . . . and you will delight in the richest of fare' (55.1–2). I wonder how thirsty for God we are this Lent? Are we simply going through the motions of what we do at this time every year? Could we come with a new expectation that God would refresh us? The Jesus who turned water into wine is with us now and delights in pouring his living water into our ordinariness for his glory.

And God said

In Chapter One we looked at the Babylonian creation story and became aware of the difference between the *Enuma Elish* warring gods and the supreme creator God of the Bible. Returning now to Genesis 1.6–8, we see the same here too. In some creation stories, it is a struggle for the gods to separate the waters into the upper and lower spaces, but not so for Yahweh: as with the creation of light, God speaks and it is so.[6] The repetition here and throughout this opening chapter of the Bible of the assertion, 'And God said, "Let there be . . ."' reinforces the power of this statement.

In *Enuma Elish*, the heavens and the earth were created out of the vanquished goddess Tiamat almost out of necessity because Marduk needed to do something with her body. By contrast, the poetic assertion that God spoke the world into being makes a very different statement. For one, this world is not simply the after-thought of a capricious god. Most importantly though, it leaves no room for chance or for life being a random occurrence: this world and the whole universe with all its billions of galaxies is something that God has willed into being, directly and purposefully.

The Genesis texts give no reason why God created the world, but we can affirm that it did not come out of some necessity within him: God is complete in and of Godself, in the fullness of the Trinity. There is nothing in the Genesis texts to indicate that God had to bring the world into being; it is simply a fact that he does, and that he does so out of his own choice.

It is worth taking some time here to explore why, theologically, the universe exists, as that will help us appreciate the different aspects of the natural world that we are considering throughout *Saying Yes to Life* and therefore our motivation to look after them.

The Mennonite theologian, Thomas Finger, talks about the redemptive activity that lies within the Trinity. He calls this 'the divine agape love' and says it is an energy that is 'always going out of itself, giving itself for another'. God is love, and love finds expression in creative generosity. From this perspective, Finger says 'it is appropriate to think of the cosmos originating from an overflow of this perichoretic *agape*. God desired that others should share in the adoration, cooperation, and joy occurring in God's own life' (we will explain what perichoretic means shortly).[7] The Franciscan writer Richard Rohr expresses the same thought when he says, 'Through the act of creation, God manifested the eternally outflowing Divine Presence into the physical and material world.'[8]

Thomas Finger goes on to connect God's overflowing, redemptive *agape* with another notion linked with redemption: that of *kenosis*. This is a Greek word that means 'emptying' or 'limiting' and is used by Paul in the beautiful hymn of Philippians 2 where he states that Jesus 'emptied himself' or 'made himself nothing' (v. 7). For Finger, drawing on a long tradition of theological thought, God's act in creation of outpouring love must also have involved an act of self-limitation. Unless we view the cosmos as something that always existed alongside God, then before the act of creation, 'God was the only reality there was'. Creation could only happen, therefore, 'if God opened up a space within herself, as it were, where this could occur. But in so doing, God would limit, and humble, Godself, allowing creatures to exist in a free space within her'.[9]

This act of redemptive limitation, seen in the incarnation, is not therefore an uncharacteristic one-off act on the part of the Godhead, but an essential part of the character of God and another way by which creation and redemption are closely intertwined. Self-limiting is also, therefore, embedded in the heart of how we conduct our own relationships.

The concept of creation existing within a space that God has allowed to open up within Godself is profoundly moving, and blows away the idea we so easily hold in our minds that we are here and God is there, somewhere else, distant to us. This is not to suggest that creation and God are one-and-the-same, and our Genesis creation narratives allow no room for any form of pantheism (literally 'all-god') that says that all things are one with God/the divine. But it also speaks clearly of the closeness between creator and created, the latter coming from the very Word of God, being sustained and held by him, and enveloped by his ceaseless love. As Bénézet Bujo, theologian from the Democratic Republic of Congo, expresses it, 'God penetrates all his creatures with his presence'.[10] This closeness allows us to speak not of pantheism but of pan*en*theism (literally 'all-in-God'). God is in all, suffusing his whole creation with his being, and all is in God. God is in everything but not everything is God. Thus the natural world is not divine, but it is sacred, 'dedicated to or associated with the divine'.[11]

One of the implications of this is that the created order is thus a reflection of who God is. The Patriarch of Romania, Patriarch Daniel, describes creation as God's fingerprints[12] and, because creation reflects God, Andrew Kyomo can look at the deforestation in his homeland of Tanzania and declare, 'by not protecting forests . . . we will be destroying the face of God whom we claim to love'.[13]

When we think of God, fundamental to Christian theology is the idea of one God in three persons, the Trinity: Father, Son and Holy Spirit. There is a word, mentioned earlier, that theologians have used since the seventh century to describe the relationship of the Godhead: *perichoresis*. This Greek term literally means 'interpenetration' and speaks of the continual movement of mutuality, reciprocity and communion that flows between Father, Son and Holy Spirit. The Celts captured this dynamism in their images of the Trinity as an interweaving triangle or circle, and the early Church Fathers imagined it as a round dance. Richard Rohr picks up on this in his language of the divine dance (and many of us will have sung Sydney Carter's 1967 song, 'Lord of the Dance').[14]

Is it any surprise, therefore, that we find relationship embedded throughout the natural world? We call those relationships ecosystems, and we know that nothing in nature exists by itself: everything exists in relationship to that which is around it and within the whole web of life. If one species is pulled out of the web, it will have knock-on consequences that can be far-reaching.

And so we affirm that the existence of this world and the universe which it inhabits has not come about by chance but through God who chose to create it out of the overflowing of the love between Father, Son and Holy Spirit. All creation therefore exists in him – lives and moves and has its being in him – in the space created within the Godhead to allow us to come to being.

Such a thing is stunningly beautiful to contemplate and can only lead to a sense of deep awe and wonder, perhaps best encapsulated in this paraphrase of Psalm 8:

O God, how full of wonder and splendor You are!
I see the reflections of Your beauty
  and hear the sounds of Your majesty
  wherever I turn.
Even the babbling of babes
  and the laughter of children
  spell out Your name in indefinable syllables.
When I gaze into star-studded skies

and attempt to comprehend the vast distances,
I contemplate in utter amazement
my Creator's concern for me.
I am dumbfounded that You
should care personally about me.
And yet You have made me in Your image.
You have called me Your child
and chosen me to be Your servant.
You have assigned to me
the fantastic responsibility of carrying on Your creative activity.
O God, how full of wonder and splendor You are![15]

Naming the river

We have looked in this chapter at God's creation of the vault between the waters, separating the waters below the sky from the waters above. Let us now start to think about water today.

In his book, *Slowly Down the Ganges*, Eric Newby describes the 1,200 mile journey he made with his wife in 1963–4, from Haridwar to the Bay of Bengal, where the Ganges-Hooghly finally enters the sea. It is a wonderful (if inevitably now out-dated) description of their journey through India, telling of the adventures they had and the colourful characters they met. It is too an exploration of the political and social currents shaping the river and the land it flows through.[16] To Hindus, the Ganges is the most sacred and venerated river on the earth. It represents the goddess Ganga and was created by the god Shiva gently lowering Ganga onto the earth (to prevent her waters descending too quickly and flooding the land) in order for her to wash over the ashes of 60,000 ancestors of an ancient king called King Sagara. The people had all been burned to death by the terrible stare of the sage Kapila who they disturbed as he meditated. If Ganga washed over them they would go to heaven. The Ganges is thus seen as a crossing point between heaven and earth: a place where heaven's blessings can most readily reach earth, and prayers and offerings are most likely to reach the gods.[17]

Newby starts his book with the 108 names by which the river Ganges is known. Some of those relate clearly to the Hindu belief in the divine origin of the Ganges, such as *Bharga-murdha-krtalaya,* 'having Bharga's (Siva's) head as an abode', and *Sangataghaugha-samani,* 'destroying the mass of sins of Sangata'. But others relate more to the physical properties of the river and the important role it plays in the lives of the people. There are names such as *Bindu-saras,* 'river made of water-drops'; *Hamsa-svarupini,* 'embodied in the forms of swans'; *Ajnana-timira-bhanu,* 'a light amid the darkness of ignorance'; *Nata-bhiti-hrt,* 'carrying away fear'; *Sankha-dundubhi-nisvana,* 'making a noise like a conch-shell and drum', and *Lila-lamghita-parvata,* 'leaping over the mountains in sport'.[18]

I find reading these and the many other names humbling and chastizing. It hardly need be said that I hold a very different understanding as to how the river Ganges came into existence, and yet I can see that giving it names has led to the river being noticed and acknowledged in a way that challenges me (though it has *not* led to the river being looked after and protected. The Ganges suffers from immense problems caused by pollution and the environmental impacts of the many hydroelectric dams that have been built.) The 108 names of the Ganges prompt me to think about the lakes and rivers in my neighbourhood and to ask myself how much I notice and pay attention to them. If I took the time to stop and reflect, what names would I give to the lake we play in so often? Or the local canal that I walk along regularly when I need a break from writing this book?

In the second creation account of Genesis 2, the human is tasked with naming the animals. We see a reflection of this in Maori culture where every aspect of the landscape inhabited is identified by name. According to Maori theologian Rob Cooper, this is not merely an act of power, it is 'the placing of such natural features within the hearts and lives of our very existence'.[19] Think about the water spaces in your area: what names might you give them and how might that help you notice and appreciate them more?

Mai ni Mwoyo

Water is a truly amazing part of our world and of what enables life to exist. Although freshwater covers less than one per cent of the earth's surface, we are utterly dependent on it for our survival and it provides the habitat for about ten per cent of the world's known species.[20] We are part of a huge hydrological cycle that uses the energy of the sun to create a constant exchange of water between the oceans, the land and the atmosphere, as water moves from the earth to the atmosphere and back to the ground and oceans, particularly through evaporation, transpiration, condensation, precipitation and runoff.[21] Water is extremely precious and the Kikuyu in Kenya say *Mai ni Mwoyo*, 'water is life', a phrase we would all do well to adopt. In traditional Kikuyu society, proverbs and taboos were developed in order to make sure people did not pollute water, and access routes and fords were always left open so no traveller could be deprived of water they needed.[22]

One striking example of the role of water in a particular ecosystem is given by German forester, Peter Wohlleben, in his fascinating book, *The Secret Network of Nature: The Delicate Balance of All Living Things*.[23] Peter talks about the rivers running inland from the north-west coast of North America. If there is one thing that these are known for, it is salmon. When they are born, far upriver, young salmon swim down into the ocean where they remain for up to four years, feeding and fattening up in preparation for their epic journey back up to the place they originated, which then becomes their spawning ground. As they swim upstream, battling the currents and the waterfalls, they have to avoid the bears that line the rivers, waiting for a good meal. For the bears it's feast-time and they gorge on the wonderful, nutrient-rich fish. As they begin to have their fill, the bears get more picky and eat less of the fish they catch, leaving the leftovers for other creatures. These are often less bold than bears and so carry the remains into the forest. The bones and head of the salmon often get left and they gradually break down and sink into the soil and fertilize it, along with the faeces from the animals enjoying their dinner.

All of this means that the forests along the river banks are extraordinarily high in nutrients, particularly nitrogen. In fact, it has been found that up to 70 per cent of the nitrogen in vegetation growing in these areas comes from the salmon – and in some trees the figure reaches 80 per cent. Insects living along the rivers can have up to 50 per cent of their nitrogen from the salmon and levels of overall biodiversity of insects, birds, plants and animals there is increased. The relationship is so close that growth rings of trees can show scientists the salmon levels in any given year, due to the presence of a particular isotope, nitrogen-15, which is found in fish and therefore also shows up in the rings. The higher the amount of nitrogen-15, the higher the levels of fish there will have been that year. Without the river and the salmon it carries, the ecosystem would be significantly poorer. Water is fundamental to its flourishing.

The hub of life

The Hungarian biochemist and Nobel Peace Prize winner, Albert Szent-Györgyi, famously described water as 'the Hub of Life. Water is its mater and matrix, mother and medium. Water is the most extraordinary substance! Practically all its properties are anomalous, which enabled life to use it as building material for its machinery. Life is water dancing to the tune of solids'.

But, today, water is both in trouble and is causing trouble. We have already mentioned the terrible degradation of the Ganges, which is the most populated river basin in the world and is suffering from industrial pollution, the impact of dams and too much water being taken out, predominantly for agriculture. This has serious consequences for the health of the 650 million people who dwell in its regions, and also for the wildlife living in and depending on it. More than 140 species of fish, otters, gharial crocodiles, turtles and many other wonderful creatures are being threatened by the degradation of the river. One of those creatures is the Ganges River dolphin which is seriously at risk: there used to be tens of thousands in the Ganges and now there are only about 15,000 left.[24]

The Ganges is not the only freshwater ecosystem in trouble. Around the world, our lakes, rivers and wetlands are among the most threatened habitats, and this has a serious impact on biodiversity. In Madagascar, 43 per cent of its freshwater species are threatened with extinction.[25] Even in the UK, over-extraction is leaving water levels too low to maintain wildlife populations. Figures for England from the Joint Nature Conservation Committee have shown that 68 per cent of rivers that are designated as Sites of Special Scientific Interest (SSIs) are in a bad condition, leading to a dramatic decline in wildlife.[26] In the twentieth century, freshwater fish have had the highest extinction rate worldwide among vertebrates and, overall, freshwater species numbers have seen an 83 per cent decline since 1970.

An 83 per cent decline means that, in the last fifty years, eight out of every ten freshwater species has been wiped out! Could you stop for a moment to allow that figure to sink in and consider how this relates to our faith in a God who made this world to be teeming with life?

Alongside the impact on biodiversity, lack of access to clean water is one of the biggest issues facing the human population today. Managing without proper access to clean water is incredibly tough. Ungwa Sangani lives with her three children, aged 18, 14 and 10, in Lulindais, a village in South Kivu in the east of the Democratic Republic of Congo (DRC). In the past, the nearest water source was a river some distance away, to which the women of the village would make a two-hour round-trip first thing in the morning, and again in the afternoon. 'I would have to leave home at four or five in the morning to fetch water to drink,' says Ungwa. 'I left so early so that I didn't meet anyone else there. We thought that if there was no one there washing clothes, the water would be okay to drink, so that is why I went so early. But we had a serious problem with sicknesses like diarrhoea, typhoid, and fevers.' Ungwa is a single parent who earns money from the produce from her fields, but the water problems gave her little time for her business. 'I would have to stop work early and leave my field to go and collect water, because it would take two hours to go to the river, collect water and take it home.'

The contaminated water Ungwa collected had to meet all her family's needs: drinking, cooking, washing, scrubbing plates and any other cleaning Ungwa could manage. But it often wasn't enough. 'Washing clothes was a problem,' she says, and 'the children were often sick.' Defecation in Lulinda took place mostly in the bush surrounding the village; the few latrines were dirty and poorly maintained; rubbish was left to rot around the village; and hand-washing with soap or ash wasn't practised. Diarrhoea and other water-borne diseases were rife, with children suffering in particular.[27] Such illnesses exacerbate the problems poor communities are facing.

Sickness affects adults' ability to work, reducing household incomes, and means that children are sometimes too unwell to attend school. Moreover, buying medicines to treat these diseases is an additional cost, and while providing a short-term solution to health problems, fails to address the cause, so people will inevitably become sick again. These conditions perpetuate the poverty that communities like Ungwa's are living in.

It is the combination of poor quality water supplies and inadequate sanitation facilities that can wreak such havoc. This is why Tearfund takes a holistic approach to water programming, considering the other crucial linked issues of sanitation and hygiene, so as to ensure that supplying clean water is not tackled in isolation and undermined by other issues.

Currently 60 per cent of the world's population (4.5 billion people) live in areas of water stress,[28] where the amount of water available cannot meet the demand in a sustainable way, and this looks set to worsen as demand for water around the world is predicted to rise by up to 50 per cent by 2050. By then, it is thought that 6.3 billion people will live in water-stressed areas, and 80 per cent of them will be living in developing countries. The demand for water is increasing more quickly than the growth in population alone, due to rising consumption, urbanization and energy needs. By far the largest share of water usage (nearly 70 per cent) is taken up by agriculture.[29]

This matters because a lack of clean water and inadequate education around hygiene leads to a range of problems. Many tropical diseases and half of all malnutrition cases are linked to these, with children not developing properly both physically and mentally. On a social level, women and children are most impacted by not having clean water or a decent toilet. It is costly in terms of time, education and income-potential to have to walk so many miles to get water for everyday household needs. And then there is the issue of safety and the risk of being sexually harassed while walking. Sometimes women might be watched by men when they have to toilet in public and, if they have – uncomfortably – held their needs in all day, they risk being attacked at night. Not having what so many of us take for granted 'can result in lost potential and dignity, ill health and even death.'[30]

Up to my knees in water

Shortages, however, are not the only water issues that poor communities around the world are facing. One of the most vivid memories relayed to me by my friend and colleague, Paul, is of a visit he made to a poor coastal community in Bangladesh. The people were living in basic shacks on a very narrow strip of land between the sea and commercial shrimp farms behind them. Their homes were largely constructed on mud and the community were continually building dykes to keep the sea at bay. Paul chatted to one woman who pointed far out and told him that was where her first home was, and that she had had six homes before the one they were now in, each moving back as the sea had risen and the coast crumbled. She was incredibly empowered and kept telling Paul, 'We don't need any help from outsiders, we can solve our own problems. The only thing we need is solid ground beneath our feet. Just give us that and we will do the rest.'

While they spoke in her home, the water was already around their knees. The community had nowhere else to go because of the shrimp farms behind them. They had retreated as far as they could and once they lost that last strip of land, they would have no choice but to leave, probably moving to slums in the cities to look for work. This was almost ten years ago. Presumably, the land has now gone and the community has dispersed.

We cannot talk about water today without looking at the impact the current climate crisis is having, and will continue to have: 'Earth's changing climate will affect the world's water supply in many far-reaching ways. It will influence water temperatures, weather systems and the amount of water in streams, rivers and aquifers. Changes in the world's water – how much, where and when it is available – are a matter of universal concern.'[31] A 2016 World Bank report predicted that as well as exacerbating already perilous situations, such as in the Middle East and the Sahel, climate change could cause water crises where there are currently none (for example in Central Africa and East Asia), impacting economic growth and pushing people back into poverty.[32]

Flooding due to rising sea levels and extreme weather events such as increased rainfall, typhoons and cyclones is worsening and, overall, floods have affected more people than any other type of disaster so far this century.[33] Floods cause many problems, not only the immediate issues of potential loss of life and the devastation of homes and businesses, but also longer-term plights, as crops and livestock are destroyed. Food becomes scarce and diseases start to spread; poor households exhaust their savings, becoming more vulnerable to the next disaster. Infrastructure such as roads, bridges, powerplants, schools and health centres are often badly damaged and livelihoods ruined, taking years to build up again. And the psychological impact of losing so much, including loved ones, and being so vulnerable can last a lifetime. In Australia, floods are the most expensive of all the natural disasters.[34]

In 2019, Cyclone Idai hit Mozambique. Graça Machel, the former first lady, declared its capital, Beira, the first city to be completely devastated by climate change. Increasingly warm air (which holds more water than cold meaning more rain falls in a shorter space of time); a drought which had left the land dry and so unable to absorb the rainfall (leading to increased run-off), and a rise in sea levels made the city extremely vulnerable. In addition, deforestation (an issue we will look at in our next chapter) meant that the floods rushed through the denuded soil and formed an inland sea.[35]

Alongside floods, drought is becoming an increasing reality for people the world over, from Australia to China to North America to Africa. At the beginning of 2018, officials announced that Cape Town, a city of four million people, would run out of water in three months. Suddenly the population had to reduce their water usage to fifty litres a day (the average person in the UK uses about three times that amount). A friend of mine, who lives in Cape Town, told me about the changes he and his family made: 'We stopped using tap water for watering the garden, no washing of cars, and having cold showers using the water that first comes out. Loads of us used large buckets to shower in and then used the collected water either to flush the loo or water the garden, and we had a grey water system installed using shower and bathroom sink water and washing machine water. "When it's yellow let it mellow, when it's brown flush it down" became Cape Town's mantra.'

How did the churches respond? The city met with church leaders and started to plan how they could help deliver water to the elderly and disabled who would not be able to walk to collection points or carry a 25 litre bucket. Church leaders also put together a peace plan to mitigate against large scale unrest at water distribution points. The Revd Rachel Mash, Canon for the Environment at the Anglican Church of Southern Africa, told me:

> We realized there are 722 verses in the Bible that talk of water. We prepared Lenten materials about water and the young people wrote daily reflections on social media. We had to learn again that water is sacred. It may come from the municipality, but they merely clean it and deliver it. Water is a free gift from God.

In Cape Town, the situation spurred people into positive action and disaster was averted, though one year's bad rain could take them back to crisis again. And along with crisis comes the recognition that Cape Town is still a city of huge inequality and many Capetonians have always lived, and continue to live, at day zero every day. They are housed in shacks, collect water from a communal tap, and have to expose themselves to robbery, rape and/or murder when they use communal toilets at night in unsafe areas. As a Tearfund colleague who lives in Cape Town said to me,

> The reality of inequality and poverty in Africa is that whilst many people are worried about losing precious environmental services, there are hundreds of millions of others who have still never experienced them. We need to worry about environmental crisis, but we also need to engage with the poverty and inequality that already denies water to many citizens, meaning they live in a permanent state of crisis.[36]

As I write this, India is in the grip of a terrible drought and globally, millions of people suffered drought in 2018: three million in Kenya; 2.2 million in Afghanistan; 2.5 million in Central America . . . the list goes on.[37] These are overwhelming numbers, but behind them are individuals struggling to survive; struggling to grow their crops to feed their families and make a living; struggling to feed their animals; struggling to keep clean.

One such person is Jumana who lives in Chad. Chad has been suffering the impacts of a changing climate, which has meant that the rains have become unpredictable and there have been devastating food crises on and off for years. Jumana has five children and has already lost one child and her father to hunger. There have been times when Jumana has resorted to digging through ants' nests in 50 degrees of heat to collect seeds buried there, taking them home for her children to eat. Every mum in her village has done that at some point.[38]

Saving water

It does not have to be this way, and we can make a difference through the actions we take in our lives and in our churches.

First, we can pray. Water is something we use often throughout the day. Why not say a prayer every time you use water in the coming week: when you turn on a tap, flush the toilet, have a shower, put on the washing machine or dishwasher? If you have easy-to-reach, clean water and don't bat an eyelid at putting it down the toilet, ask God to help you remember those who are in a different situation.

Second, we can give. On my bathroom wall, next to the loo, I have a picture of a toilet in Bihar, India. It's very basic, but it is clean and decent. I asked to have the bathroom toilet twinned for Christmas one year, and it is nice to know that the money spent on my Christmas present has gone to provide a community with a proper toilet, plus the hygiene education process that accompanies it.[39]

Many organizations are involved in helping provide proper access to water and sanitation facilities, and training around how to use them well. In development speak, that is called WASH – water, sanitation and hygiene – and is encapsulated in Sustainable Development Goal 6, 'To ensure availability and sustainable management of water and sanitation for all [by 2030]').

Let's return to Ungwa in the DRC. She is now in a very different situation because of work being done by the partnership (already mentioned) between the ODI (Overseas Development Institute), Oxfam and Tearfund. The community now has access to clean, safe water in the heart of the village. Many families have constructed latrines with 'tippy-taps' where they can wash their hands, and pits where they can get rid of their rubbish. Hygiene has improved considerably and there is now a 'healthy village' committee with a water sub-committee. Ungwa has seen a significant improvement in the health of her children: 'I've noticed that my children are less sick and we don't have to go to the clinic like we did before,' she says, and the health benefits of the process have been felt across the whole community, saving people money and giving them more time for their families and livelihoods.[40]

Third, we can take practical action. Think about the water in your area. Maybe there are local initiatives you could get involved with that look after your rivers or canals. Could you or your church join in or initiate a clean-up? A Rocha UK was started because Dave and Ann Bookless saw a large area of abandoned open space near them known as 'the Minet tip', which included a small river, clogged up with rubbish. Over five years, with the involvement of the community and local churches, they transformed it into the Minet Country Park. One of the highlights has been the clearing of the river itself, which now attracts all sorts of wildlife including kingfishers, newts and dragonflies.[41]

The Church can also be present in standing up for the protection of waterways and the people and ecosystems that depend on them. US Episcopalian, Bishop Michael Curry, in his Letter for Creation to Justin Welby, talked of the role of the Church in the fight to stop the Dakota pipeline which would have split the Missouri river and thus put at risk that vital source of water for the Standing Rock Sioux's reservation:

> I saw people of every nation, faith, age and race move to stand with the Standing Rock Sioux as they struggled to turn back a pipeline that threatened their sacred lands and their water supply. And I saw the Episcopal Church flag at the front of that procession. When crowds chanted 'Mni Wiconi' (water is life),

Episcopalians chanted with full voice because we have been given new life in Jesus Christ through the waters of baptism. Yes, water is life. Yes, we should honor it.[42]

In our own lives, there is practical action we can take to reduce our water usage, whether we live in places where doing so really matters, or whether we choose to participate as an act of solidarity. We can take short showers; turn off the tap when brushing our teeth; only run the dishwasher or washing machine when it is full; install water butts in the garden and use those for watering; and allow our car to get dirty! One thing to bear in mind is that our biggest water usage actually lies in the things we consume, our 'virtual water'. For example, a pair of cotton jeans uses 9,500 litres of water to produce, and a beef burger uses about 1,000 litres (in comparison to a soy burger which uses just 160 litres). So, the key way to use less water is to consume fewer things.[43]

We can help reduce water pollution by becoming more aware of what we are flushing down our toilets and sinks and changing that. There are vast amounts of chemicals in cleaning and personal care products, so take a look in your kitchen and bathroom cupboards: is there one product in each that you could switch for an environmentally-friendly version?[44] Also, intensive agriculture is responsible for a lot of the pollution in our waterways as the chemical inputs that are used then soak into the ground and run-off into rivers, etc. So, take steps to support farming practices that are reducing chemical usage or are fully organic. Buying from local farm schemes enables you to talk to the producers and find out more about their practices. You could always try growing your own fruit and veg, which can be a cheaper option.

We have seen that climate breakdown lies behind many of the problems that are facing water today, so let me reinforce the message of the previous chapter. If we do not act boldly towards getting our CO_2 emissions to net zero and keeping within 1.5° of warming, floods and droughts will get worse and suffering will increase. Have you taken any of the steps talked about at the end of Chapter One? How has it been? What else could you do?

Unending life

The nineteenth-century hymn by George Cooper has us singing, 'The Spirit is a fountain clear, forever leaping to the sky; whose waters give unending life, whose timeless source is never dry'.

In this chapter we have looked at the theme of water in the Scriptures and then at water today, seeing both its beauty and its problems. I hope you have been inspired to marvel afresh at water, to notice where and how it features in our lives; to appreciate its wonder, and commit yourselves to taking care of this precious resource that is so vital if life on earth is to flourish.

May we commit ourselves again to Jesus, the Water of Life. May we ask him to fill us with the clear fountain of his Spirit, and as he gives us unending life, may we live and speak in ways that pass that life on to all that is around us.

For discussion

1 Look back at the names given to the River Ganges. Think about a favourite stretch of water and consider what names might help you to acknowledge and pay attention to it.
2 In this chapter we have seen that, 'the existence of this world and the universe which it inhabits is not by chance, but comes from God who chose to create it out of the overflowing of the love between Father, Son and Holy Spirit. All creation therefore exists in him – lives and moves and has its being in him – in the space created within the Godhead to allow us to come to being.' How does this expand your understanding of this world and its relationship to God?
3 We know that Jesus was baptized in the River Jordan, but do you know where the water that is used for baptism in your own home church comes from, or what processes it has gone through to get there? How might you protect your own River Jordan?
4 What personal action will you take to look after the world's precious resource of water and those that depend on it?
5 In this chapter's interview, the Archbishop of Cape Town, Thabo Makgoba, reflects spiritually and practically on water. Watch at

<www.spckpublishing.co.uk/saying-yes-resources> and use the interview in your thoughts and discussions.

A prayer on water from Nigeria

Dear Lord, it is exciting to know that you are the very source of life including water. We praise and adore you for the gift of water that sustains all life and constantly reminds us that You are the fountain of living water.

Teach us to use it thankfully, to consume it consciously, and to protect its purity.

Father, forgive us for the times we took it for granted. We confess our attitudes of greed, dominance, and insensitivity towards your beautiful creation, and particularly towards water. Lord, forgive us for the times we have used water selfishly, unwisely, and without regard for how it affects others. Forgive us for the actions we have taken to harm the different sources of water around us.

Help us to see the effects of our actions not only on our immediate surroundings but also on people living in places plagued by drought. Help us to be conscious of our daily use of water; help us to be more willing to reflect on its symbolic nature and the lessons it teaches about you and your sustaining power.

Please guide us on how to protect the water bodies you made for your glory. Amen.

Prayer by Fwangmun Oscar Danladi. Oscar is a youth pastor at the ECWA Good News Church, Jos, and social activist at the Jos Green Centre, a centre for eco-entrepreneurship for young people.

3
Let the land produce vegetation (Genesis 1.9–13)

> [9]And God said, 'Let the water under the sky be gathered to one place, and let dry ground appear.' And it was so. [10]God called the dry ground 'land,' and the gathered waters he called 'seas.' And God saw that it was good.
>
> [11]Then God said, 'Let the land produce vegetation: seed-bearing plants and trees on the land that bear fruit with seed in it, according to their various kinds.' And it was so. [12]The land produced vegetation: plants bearing seed according to their kinds and trees bearing fruit with seed in it according to their kinds. And God saw that it was good. [13]And there was evening, and there was morning—the third day.

The highlands of Cusco, Peru are one of the regions of the planet that has been severely affected by climate change.[1] Talk to any Quechua farmer and they will tell you how things have altered in recent decades: the rains are no longer reliable; there are unusual droughts called *veranillos* during the growing season; changes in temperature mean crops have to be grown at an increasing altitude, and there are more unpredictable events, such as hailstorms, that destroy crops.

'Before, we knew when rains were to start and to end during the year. This helped us in our farming. But that's no longer so. Crops don't produce well. The climate has changed', says one Cusco rural farmer. Victor, another farmer, puts it this way: 'These days, seeds don't grow as they did before; the day isn't long enough to finish our work. And then . . . everything is expensive'.

Trees have a vital, multi-faceted role in countering climate change and mitigating its effects. They remove carbon dioxide from the atmosphere, hold back storms and flooding, and help protect and restore moisture and fertility to soils, improving agricultural conditions. They also provide vital habitats for other species and serve as barriers to protect high Andean crops, as well as livestock, from potentially harmful frosts, hail and strong winds. Yet, many of Peru's mountains have long been denuded of native trees as Quechua communities have used the wood for construction and cooking.

A Colombian friend called Juliana has lived in Cusco for many years. She is absolutely dedicated to helping churches understand the biblical call to appreciate and care for God's diverse creation. As well as teaching, she also looks to get churches, and especially Christian young people, involved in practical environmental initiatives that can enable them grow in understanding and commitment and be a witness in their local communities. Her work is being supported by a Baptist church in Guildford in the south of England.

One of those initiatives took place high in the Sacred Valley, and Juliana described it to me like this: 'Just imagine . . . 350 people from three rural, highland Quechua communities, dressed in bright red traditional clothing and joined by a number of volunteers, each carrying between fifty and one hundred queuña saplings to plant, marching in a line up the mountain. We climb and climb . . . and then together, two by two, carry out this massive tree-planting task. This initiative, called *Queuña Raymi* (literally, Queuña Festival) is a true celebration.'

The day led to 32,000 native *queuña* trees being planted in the Rumira Sondormayo community, at 13,800 feet above sea level, and it is part of a project to form a forest of a million trees that will cover the bare mountains and replenish the watersheds below.

Juliana went on to tell me, 'Many of the Quechua families involved are Christians. They were so surprised and appreciative that Christians are supporting them in this initiative. Besides reducing the carbon footprint and generating oxygen, they have also increased local water reserves, while other surrounding regions are experiencing increased water shortages.'

Another initiative saw Juliana working with a church in Cusco that had a vision for a prayer mountain. Literally: it had bought a mountain on the outskirts of the city (in earlier years when land was cheap) to turn it into a 'mountain of prayer for the nations'. But the leader, Pastor Americo, wanted to combine prayer with the restoration of the land and so, with Juliana's help, 40 volunteers from different denominations in the city joined forces to plant 1,000 trees.

Juliana tells me that, 'in this way, the church gave a vibrant Christian testimony in their community and, thousands of miles away, in the UK, another church showed its concern and sense of responsibility for how climate breakdown is affecting these communities, by sponsoring these initiatives.'[2]

Land and seas

As we continue following the story of creation in Genesis 1 through this Lenten time, we see a beautiful movement taking place. We started with the emptiness and formlessness of watery chaos, and as the Spirit of God moved over it, so God spoke and light was created, preparing the conditions for life and for his creation to be seen. Then order was brought to the waters as they were separated into the water in the atmosphere and the water on the earth. Now we see the waters on the earth being gathered together and drawn back like a curtain so that dry ground might appear. As that happens, so the land and the seas are brought into being. Can you envisage this in your mind?

In Chapter Five we will consider the seas in more detail; in this chapter our focus will be on land and trees, looking both at their roles in the great story of salvation in the Bible and then at how they feature in our contemporary context. We live in a beautiful world, with many of us surrounded by green – I see green whether I look to the left out of my back door, or to the right out of my kitchen window at the front. My hope is that we can be inspired to a new love of the land and what grows on it and, as we move this Lent towards the death and resurrection of the Lord of all creation, to see what place they hold in our Christian faith.

On this third day in Genesis 3.9–13, we continue with the creation of the spaces and the environments which the created beings of Days Four to Six will then inhabit. Having fashioned the seas and the land,

God pronounces that the land should produce plants and trees. It is beautiful language: 'Let the earth grow green [with] grass, plant yielding seed, fruit tree bearing fruit, according to its kind [and] which has seed in it on the earth.'[3] Life thus emerges in order and symmetry, like a flower unfolding, opening up and revealing intricacy and loveliness.

In these verses, fertility is blessed and made a part of God's world. As we will see later in this book, God's creation is abundant – this is no miserly God being described here! This is a God who loves to bless; a God who delights in growth and richness, who wants his creatures to live in fullness of life. I have learnt that one of the dilemmas of growing my own vegetables from seed is that they often produce far more than I need. The tomato seeds give me a trayful of seedlings when I only want a handful for myself, so I enjoy giving most of them away, and the beetroot or salad seeds I sow come up thickly, so rather than let them grow to maturity, most of them go into a salad or stir-fry as I thin them out. Of course, there are reasons why plants produce so many seeds, and in the wild most of what is produced will meet too many hazards for all to grow. Yet, these verses speak of a fecundity that lies at the heart of a God who is three-persons-in-one, and who wants always to be giving to the other in generosity and overflowing love.

And God saw that it was good

In our Day Three verses we see the second and third occurrences of the statement that 'God saw that it was good'. This phrase is used seven times in Genesis 1, including v. 31 where God declares that all he has made is 'very good', which reflects the understanding of completeness that the number seven holds within Judaism.

The importance of this pronouncement of the individual aspects of creation as being good should not be underestimated. It has not always been understood in the Church. In fact, the American nineteenth-century preacher, D. L. Moody, said famously about his calling, 'I look upon this world as a wrecked vessel. God has given me a lifeboat and said to me, "Moody, save all you can".' His view has been very influential within Christianity, contributing to an understanding of the Christian faith that says the created order is doomed to destruction and our mission is to save souls onto the lifeboat of the Church and whisk them off to salvation in heaven. I have had the privilege of teaching on the Bible and environmental care in many different contexts, involving church leaders from around the world, and I often meet people who tell me that the Church's job is to 'preach the gospel' and plant churches, and anything else is a distraction. I encounter these attitudes online too, as in the tweet I received from someone telling me, 'With the Church in the state of collapse that it is, you should be planting churches not trees'!

In the next chapter we will look more closely at different Christian views on the future: what will happen to the world (indeed to the whole universe[4]) and whether our final resting place will include the earth. The key point here is that the theological understanding we see in Moody's deeply-held belief – and in the views of those who consider the role of the Church as exclusively 'preaching the gospel' and planting churches – is rooted in a negative view of the world. In previous chapters we explored how the Genesis text depicts the world as coming into being, not through the actions of squabbling gods, but due to the one true God deliberately choosing life. Reflecting what we saw in Chapter Two, former Archbishop of Canterbury Rowan Williams draws on the Russian Orthodox theologian Father Sergei Bulgakov, and on Jewish mysticism of the late Middle Ages, in talking of creation as having come from the loving God making space for it through his creative and generous heart. God is creator, not only at the moment when the act of creation occurs, but in his very nature. And so the eternal nature of God becomes visible in his creation: 'Creation translates into time and limit and history the eternal fact of God'.[5]

How then can something that God has declared good; that comes from the overflow of his love; that is the 'translation' of who God is, and that is continually sustained by his Spirit be described as a sinking vessel?

British New Testament scholar, Richard Bauckham, points out that the declaration 'it was good' at the end of each day in Genesis 1 (as opposed to only at the very end when all was completed) indicates that 'each part of creation has its own value that does not depend on its value for other parts. The environments . . . are not valued only because they serve as environments for their inhabitants'. Whilst the provision of food and habitat is of course a key part of what is being described in Genesis 1, yet 'God appreciates the trees and plants also for their own sake'.[6] What a beautifully simple yet profound statement!

The astrophysicist and theologian, Professor David Wilkinson says, 'matter matters to God'. It is a wonderful phrase and one that destroys a dualism that has its roots in pagan Greek Platonic thinking, but has become a prevalent part of Christian theology. This dualism separates out body and spirit, earth and heaven, natural and spiritual. It exalts the latter and denigrates the former, so that nature/creation is held to be inferior to the 'supernatural' realm.[7] We see this dualism at play in many aspects of the church. We encounter it when we describe church leaders as being in 'full-time Christian ministry', rather than viewing all of us as working full-time for God, whatever sphere of life we are in. We see it when we separate evangelism from acts of practical care, and when we restrict our worship to something that we sing or say on a Sunday. It creeps in when we talk about 'saving souls', and it's there again when we declare this world to be of less value to God than heaven and as something that will be destroyed. We see it when we sing hymns with lyrics like, 'This world is not my home, I'm just a-passing through'.

The world has paid a price for this dualism. Swedish theologian Mika Vähäkangas, reflecting on the environmental crisis in Tanzania where he lived, comments on the sacred/secular dualism that Protestant missionary theologians brought with them to Africa, and how alien this was to African traditions, as well as being unbiblical (as we will see further in Chapter Four). He believes that 'a major reason for the environmental crisis today is the way the western thinking has demystified nature and included it in the sphere of the secular'.[8] By contrast, as highlighted in the previous chapter, God's appreciation of his creation as 'good' allows us to see the land with its plants and trees as sacred. Maybe we can learn something from Native American

Cheyenne priests who touch the earth four times before a ceremony. This is done in recognition that the earth is a creation of one creator God. It is to be acknowledged and not ignored.[9]

And so God declares the seas and the land, the plants and trees to be 'good', prompting Bonhoeffer to talk about 'the profound this-worldliness of Christianity'.[10] But let us notice that this declaration comes not as an abstract assertion, but as God's response to looking on what he has made. It is, as American theologian Ellen Davis puts it, 'a divine perception'.[11] In environmental ethics there is debate over where the value of nature lies. Is it extrinsic (does its value lie outside of itself as a resource for us to use, as seen in the language around 'ecosystem services'?) or intrinsic (does it have value in and of itself, regardless of its value to people?) The Genesis narrative tells us that the value of the seas and the land and the trees and of all created things (including people of course) lies in God and in his perception: as I have often heard my friend, Dr Dave Bookless, Theology Director for A Rocha International, say: the value of nature is *theocentric*.

To know that God sees what he has created and, in his seeing, gives value to it and pronounces it to be good, is a deep call to us, similarly, to reflect this Lent on how we view the world around us. If we are really honest, do we perceive the wider world primarily as a resource for us to use as we wish, or do we regard it primarily as something that God loves and is precious to him? In *L is for Lifestyle: Christian living that doesn't cost the earth,* I talk about a beautiful tapestry I made some years ago that hangs proudly on my wall: a William Morris design of a peacock in the woods. Can you imagine how I would feel if I came home one day to find my daughters had put it on the floor and were using it to wipe their muddy feet? I'd be horrified . . . devastated . . . so upset! Well, I can tell you with complete certainty that would never happen. Why? Because they love me and would never dream of doing something so terrible to the tapestry I value so much. And they love it too, because I love it. God's affirmation of the goodness and value of this world (not a sinking vessel) spurs us on therefore to take care that we do not wipe our ecological footprints all over it, leaving it damaged and wrecked.[12]

Living in the land

The affirmation of the goodness of the land and the plants and trees is a reminder to us that we do not have a disembodied faith but one that is rooted: rooted in place and land and in the whole world that God has created. The wider creation is not simply the background – it is the context within which we live out our faith and is an integral part of how we work out our salvation.

Genesis 1.9–13, particularly its emphasis on seed-bearing plants and trees,[13] reminds us that the people of God in the Scriptures lived their lives deeply dependent on agriculture. And it is a reminder too that, however urban we may have become globally, we all depend on agriculture for our existence, even if that connection may feel distant for some of us and is easily forgotten.

The land thus plays a crucial role in the story of God's people in the Old Testament, and how they live on it is a key part of how they walk with God and follow his ways. They are always to remember that the land belongs ultimately to God: 'The earth is the Lord's and everything in it' (Ps. 24.1). Moreover, the particular land they have been given is exactly that, a gift, promised right back in the covenant God made with Abram in Genesis 12, and one which they must not take for granted or abuse. They are to see themselves as tenants of the land (Lev. 25.23). As Moses tells the people, 'You may say to yourself, "My power and the strength of my hands have produced this wealth for me". But remember the Lord your God, for it is he who gives you the ability to produce wealth, and so confirms his covenant, which he swore to your ancestors, as it is today' (Deut. 8.17–18).

Many of us are used to reading the Old Testament as the story of God working to restore his people's relationship with him, through the laws he gives and the establishment of the priesthood and the sacrificial system. And of course that is right. But it cannot ever be divorced from how the people relate to God's creation, both human and wider: the Old Testament is the story of a chosen people *in a promised land*. The quality of the people's relationship with God – their righteousness – is seen precisely in how they treat one another (for example, whether they take care of people or perpetuate injustice and oppression, e.g. Isaiah 58.6–7), and how they treat the land and other creatures (for example, whether they allow it and their animals to have Sabbath rest or flog them mercilessly to produce more and more, e.g. Lev. 25.1–5).

The land itself does not stay inert and silent in the background and the Scriptures portray it as having an agency of its own. Psalm 148 shows all manner of God's creation exuberantly praising God. This includes not only living creatures but aspects of the land too, animate and inanimate: the waters above the skies; the mountains and hills; fruit trees and cedars. Isaiah describes the mountains and hills as bursting into song before their God on account of the people returning to him, and the trees of the field clapping their hands (55.12). Looking at the New Testament, we see Paul describing the whole creation as groaning in pain as if in labour, eagerly waiting for the children of God to be revealed. The New Testament scholar, J. B. Phillips, has translated this as 'standing on tiptoe', a lovely image of impatient anticipation (Rom. 8.19–22).

Paul's depiction of the creation groaning picks up on the sad reality that, alongside praise, the land also responds negatively as it witnesses the sins of the people (as they fail to practice justice for and care of the needy in their midst), and the judgment that God brings on them as a result. Both Isaiah and Jeremiah describe the land as mourning (Isa. 24.4, Jer. 4.28), and Joel speaks similarly of the ground mourning (1.10).[14]

So the land – in all its dusty, soily, muddy, earthy physicality – is an indispensable part of the story of salvation, and it is no surprise therefore that the future (which we shall explore more in the next chapter) retains that dimension. The Old Testament prophets envisage a time when the people will live in their own houses and eat the fruit of their own vineyards (Isaiah 65.21), and the New Testament looks further ahead – as we have seen already – to a picture of a garden city with a river and trees.

The concept of the garden city may give hope to those of us reading this who are urbanites and wondering how we got into all this talk of soil and land, when our experience is primarily of concrete and glass! The Bible reflects diverse contexts and the people of God – particularly in the New Testament – are to be found in the cities as well as in the countryside. Cities can be good: they may provide safety and, through our creativity in urban design, we can reflect the creativity of God. But Genesis 1.9–13 brings us back to the essential reality that even in the midst of the most intensely urban setting, we need the light, water, land and seed-bearing vegetation that God has created in order to survive, and we neglect to think about those things at our peril. Beyond mere survival, there is plenty of evidence to show that having nature areas both within and around our cities brings a wealth of health and well-being benefits to us as well as, of course, aiding biodiversity and therefore other species.[15] So whether urban or rural, we need to be looking after the land.

Acknowledgement of country

In this respect we have much to learn from those who have traditionally lived closer to the land than many of us. The Venerable Karen Kime is Archdeacon for Indigenous Ministries with the Anglican Diocese of Canberra and Goulburn in Australia and is herself a Birripi woman. In a speech she gave to the Victorian Council of Churches she began with an 'Acknowledgement of Country', saying,

> 'I'd like to acknowledge that we are in Wurundjeri Country. We give thanks for its beautiful borders of mountains and streams and the way in which it continues to provide for her people. We acknowledge the many Elders and Custodians who continue to care for this land and who are the knowledge keepers and leaders of their communities. We also give thanks for the many Aboriginal people and families who call this place home'.

She went on to talk about the rich and diverse nations of south eastern Australia:

> *Barkandji Country* – the place of hot and arid plains, where one can see the footprints of Biamee, the Creator Spirit of the Barkandhi people.
>
> The *Yuin nation* on the south coast of New South Wales, whose stories surround the sea and who jointly manage the Booderee National Park. The Yuin people pay attention to the skies and

> the presence of the sea eagle, known to represent the 'father and protector' of them all.
>
> The *Wiradjuri people* refer to themselves as the River People and are deeply attached to the rivers that flow through their Country. The Wiradjuri have a wonderful saying that one needs to be 'calm like the water, and strong like the current'.
>
> The *Ngungawal people* – the people of the highlands whose stories and Country include the foothills of the snowy mountains and whose Country was an important meeting place for the many surrounding nations.[16]

This challenges me. In all my years of attending conferences, Christian festivals and church services, I am not sure I have ever heard anyone start their talk by acknowledging the area, its natural features and the people who live there in the way Karen Kime does, nor have I ever done so myself. It is a practice I shall learn from.

I recently met a woman called Jocabed Reina Solano Miselis, who is from the Gunadule people and was brought up on the Guna Yala islands off the coast of Panama. One of the names given by the Guna to the earth is *Nabgwana*, which means 'the mother who provides abundant fruit and expresses intimate sadness and joy in the beings created within her'.[17] She writes of the Guna practice of burying the umbilical cord and placenta in the ground when a baby is born. The women cut the cord and wrap it with the placenta and give it to the grandfather. He takes the umbilical cord and placenta to the mountain, plants a cacao tree, buries them with the tree and sings:

> Our good and great God, we thank you for the life you give this baby girl/boy. We have come from the earth and we give back to the earth. Today we bury these symbols of life and give back of your own generosity so that, just as the child grows strong and healthy, this cacao tree will grow big and strong. For we are one, humans and the earth.

What a powerful way of showing our connection to, and dependence on, the land. As Jocabed Miselis says, 'For nine months the umbilical cord and placenta united the baby and the mother. Now the cord ties men and women to the earth. It fertilizes the earth from which a plant germinates as a sign of unity and of the hope for future generations'.[18]

Maybe this sounds romantic and idealized to those of us used to a birthing system that involves the afterbirth being whisked away out of sight. I am reminded of the furore caused when the chef and campaigner, Hugh Fearnley-Whittingstall, cooked a placenta on television for a couple to eat with their family to mark the birth of their first grandchild. But I reflect also on how some Australian Aboriginal people rub earth over their children when they are born, and I wonder when you or I last actually touched the earth and felt the soil? Shane Claibourne says he resolves to 'regularly get my hands into the garden, so that when I type, I can see soil under my finger-nails', and Rowan Williams once wrote, 'Receive the world that God has given. Go for a walk. Get wet. Dig the earth'.[19] Are there practices you could develop to remind yourself of your connection with the land that you are living in and are a part of?

This loss affects everyone

We have much to learn from indigenous people groups, as Miselis says, 'not merely as part of the current fashionable trend to include those traditionally discriminated against but rather as legitimate models with hopeful proposals for the world. If we do not work on these proposals . . . we will lose a great deal as a society. With the disappearance of indigenous languages, knowledge of medicinal plants is also lost, as well as other insights into humanity's relationship with the earth. This loss affects everyone'.[20]

In reflecting on the creation of the land in Genesis 1 and its place in the story of God's people, we must also recognize the role the land plays in creating identity and therefore the deep grief that is experienced when people are robbed of their land, and the ongoing problems that ensue. We cannot speak about land today without also talking about colonialism and the impact this has had as land has been ripped from people's lives, and – through the trans-Atlantic slave trade – people were ripped from their lands. And, we must acknowledge that this is a dark seam that runs through even the history of the Old Testament as the Israelite possession of their land came from the dispossession of those who were already living there, resulting in long-term problems which we still grapple with today.

Karen Kime talks about how colonization in Australia has created 'wounds that every Aboriginal person and family continues to experience. These wounds derive from the very personal experience of inter-generational trauma. Every Aboriginal family in every community has such a story'.[21] Rob Cooper describes this as 'the bitter fruits of colonization' and talks of the cultural invasion which meant that thousands of Maori 'turned their faces to the wall during the 1800s and quietly died',[22] and Ernst Conradie talks of 'the destructive legacy of (neo)colonialism and mission in Africa'.[23]

Writing from a North American perspective, that legacy is described well by John Mohawk:

> Colonization is the greatest health risk to indigenous peoples as individuals and communities. It produces the anomie – the absence of values and sense of group purpose and identity – that underlies the deadly automobile accidents triggered by alcohol abuse. It creates the conditions of inappropriate diet which lead to an epidemic of degenerative diseases, and the moral anarchy that leads to child abuse and spousal abuse. Becoming colonized was the worst thing that could happen five centuries ago, and being colonized is the worst thing that can happen now.[24]

One of the tragedies is how linked the Church has been to colonization, and therefore to the dualism and negative view of the earth that we saw earlier. Political domination went hand-in-hand with the domination of the land as the traditional sense of kinship and connection was destroyed. Former Archbishop Desmond Tutu is well known for having said, 'When the missionaries came to Africa, they had the Bible and we had the land. They said "let us close our eyes and pray." When we opened them, we had the Bible, and they had the land'.[25] He may have meant this humorously, and of course there were many missionaries who did *not* take the land, but still some truth remains. With this recognition comes the need for repentance, both by those of the Christain faith (Protestant and Catholic) and those outside the Church who recognize the almost unfathomable damage done, and being done, to indigenous peoples as a result of colonization.

The right to land

One key issue in colonization, and when thinking about the land in general, is land rights. This is a topic that many readers in economically developed countries do not give a second thought to, and yet for billions around the world it is of the utmost importance. The region of La Mosquitia is on the east coast of Honduras.[26] Containing forests, lagoons and a rich variety of animals and plants, it is home to various indigenous peoples, whose lives are strongly linked to the land. They have traditionally used their land for gathering food, hunting, fishing and collecting wood for building canoes and houses.

Mopawi, a Christian development organization, has been working in La Mosquitia since 1985. At first it was employed on projects such as improving crop yields and accessing clean water and sanitation, but soon realized that longer-term problems were being caused by the fact that the indigenous peoples had no legal rights to the land or the forest. The most significant problem for them was land grabbing. Usually heavily armed, the grabbers – often companies being funded by foreign investors – fenced the land off and deforested it for agricultural or extractive use.

Supported by Tearfund, Mopawi started working with the communities to bring together an indigenous peoples' collective right to the land, forests and rivers. It discovered that there were no Honduran laws allowing people to claim collective land rights – especially when indigenous peoples were the ones asking. But Mopawi kept insisting that the government address the issue and over time the community members learned how to advocate and speak out for themselves, even talking to the president of Honduras himself.

It took from 1987 to 2012 to gain the first collective land rights for a cluster of 39 communities along the coast. After this, the government granted eleven more land and territorial titles, giving indigenous people the right to the natural resources as well as the land itself. In total, the amount granted is now 14,000 square kilometres.

Compared to the 30–40 million hectares of land around the world that has possibly been acquired by foreign investors to meet our demand for resources (a lot of which would fall into the land grabbing category), this amount of land is tiny.[27] But it makes a huge difference to those who live on it.

Away from La Mosquitia, in the centre of São Paulo, Brazil, families literally had to live in the dark. The state electricity company, Electopaulo, cut the power to buildings, claiming that the wiring was a fire hazard and that the supply had been switched off to keep people safe. However, land values had been rising and many residents believed the company was deliberately making the buildings uninhabitable to force families out. They tried to negotiate with Electopaulo, but their efforts were unsuccessful and they were forced to use gaslights and candles to illuminate their homes, quite a fire risk. CAFOD's partner APOIO accompanied the families in their negotiations with the authorities. After a lot of lobbying, the families were finally given the right to remain and the electricity was switched back on.

The families agreed with Electopaulo to pay for the cost of an engineer and materials to carry out the work needed to meet the fire safety standard. This was a lot of money for these very poor families to find and they are paying it off in instalments, but this demonstrates their dedication and determination to stay and how much they value the land they live on.

In South Africa, the Land Claims Commission has been established to redistribute land taken from its owners during the removals in the 1970s under apartheid. One group that has lobbied to claim back their land rights is the Roosboom United Churches Committee (RUCC) in central KwaZulu-Natal, who lost twenty churches and community members' homes under the regime. Their advocacy efforts have been very successful, and in 2017 the commission began to pay them the compensation they were owed, allowing some of the churches to start reconstructing the buildings that had been destroyed. At least five of these have now been completed, and this has meant that rather than meeting in improvized shacks, congregation members have been able to return to worshipping in a formal church building.

There remain, however, a number of outstanding claims from other churches in the community that have been subject to various delays, with the issue of land rights being deprioritized by politicians once they have secured the votes they need for a successful election. Christian Aid partner, Church Land Programme, has been working with RUCC to push for the commission to finalize these claims and restore to people the land to which they are entitled.

Trees of life

Let's return to the biblical text and the creation on the Third Day. Having drawn back the waters and created land, God then calls for the land to produce vegetation – plants and trees that bear seeds and fruit – as he continues to create the environments that will be populated in the coming Days. So from land, we now turn for the rest of this chapter to trees.

I wonder whether you have ever given much thought to trees in the Bible. Once we stop and pay attention, we notice that trees feature through the whole story of the Bible and are present at nearly every major occurrence. As Professor Julian Evans, leading forester and horticulturalist, says, 'It continues to impress me that even in such ordinary things as trees and woods we find encapsulated the elements of the gospel.'[28]

From the tree of life and the tree of knowledge of good and evil to Eve and Adam eating the forbidden fruit in the second creation narrative of Genesis 2; from the Lord meeting Abraham near the great trees of Mamre to him speaking to Moses through the burning bush; from the use of trees in the laws of purification that God gave to the Israelites (e.g. Lev. 14.49–53) to their use as a site where the Judges sat and then in the building of the temple, we see trees all the way through the biblical narrative. Solomon is described as someone with great wisdom, insight and understanding, and this extends to his knowledge of plant life, 'from the cedars of Lebanon to the hyssop that grows out of walls' (1 Kings 4.33). Trees are at the heart not only of the nation's worship of the Lord in his temple, but also at the heart of the false worship that sent them finally into exile as 'they set up sacred stones and Asherah poles on every high hill and under every spreading tree' (2 Kings 17.10).

Trees feature strongly in the Prophetic literature and are often used to speak of judgment. Julian Evans describes the cedar as being like 'the Rolls-Royce or Ferrari to add to the forecourts of the day. This lovely, precious and increasingly scarce timber was used for great works like the temple, and used by the greedy as a status symbol'.[29] So, it is with heavy irony that Ezekiel speaks of Assyria as being like a majestic cedar of Lebanon 'with beautiful branches overshadowing the forest'. Not even the cedars in the garden of Eden could compare with it and yet God cut it down to show other nations that they are all mortal and destined for death (Ez. 31).

But alongside judgment, trees speak also of hope, and particularly hope in the Messiah, the righteous branch, beautiful and glorious and bearing the fruit of the Lord (Jer. 23.5, Isa. 4.2). The people look forward to a time, as we saw above, when they will eat the fruit from their own trees and vines, and when they will not need to gather wood from the fields or cut it from the forests because they will use the weapons no longer required for warfare for fuel (Ez. 39.9–10).

One of the best-loved passages concerning trees in the Old Testament is Psalm 1. Let us hear it afresh in this version:

Those persons who choose to live significant lives
  are not going to take their cues
  from the religiously indifferent.
Nor will they conform to the crowd
  nor mouth their prejudices
  nor dote on the failures of others.
Their ultimate concern is the will of God.
They make their daily decisions in respect to such.
Compare them to a sturdy tree
  planted in rich, moist soil.
As the tree yields fruit,
  so their lives manifest blessing for others
  and are purposeful and productive.[30]

The soil is moist because, as most translations render it, the tree is 'planted by streams of water'. It is a lovely image, reflected by Jeremiah too who describes the one who trusts in the Lord as 'like a tree planted by the water that sends out its roots by the stream: it does not fear when heat comes, its leaves are always green. It has no worries in a year of drought and never fails to bear fruit' (Jer. 17.7–8).

I am confident this is the yearning of all our hearts, to be firmly planted with our roots going down deep into the Lord who created water and who is, himself, the water of life. In *Just Living*, I write of the need for us to build rhythms of space and silence into our lives: gaps when we can stop, be still, reflect and simply be. Creating such a rhythm is, I believe, a key way of building resilience and learning both how to resist the never-ending demands of our consumer culture and finding sustenance for a journey that is often tough and wearying. I talk of my own practice of silent meditation and how I sometimes visualize a cross-section of a river. The flow of the river runs along the top, representing all my thoughts that course along constantly. When I am aware of particular notions coming into my head, I can consciously throw them into that river of my life and into God's care. At the bottom of the river is the bed where the rocks lie motionless, and it is there, at that still point, where I meet with God when I am sitting in silence.[31]

This is a slightly different image to the one the psalmist talks about, but I wonder whether you could use it to help you root yourself into the stream of living water that God so graciously and generously provides for us? In this Lenten period, as you seek time to reflect and pray and let go of some particular habits, are there new practices you could take up to help you be like that tree planted in rich, moist soil?

I am the vine

As the history of God's people and his wider creation unfolds, we see trees are constantly present, their branches spreading around the whole story. This is no less true in the story of Jesus' life in the gospels. We associate Jesus' birth with sheep and stars, but trees are there too in the gifts the Magi bring. Both frankincense and myrrh are resins that come from the sap of their respective trees, which is collected by cutting the bark so the tree 'bleeds'. Both resins can be burnt to release an aromatic fragrance, and frankincense can also be made into oil. Neither frankincense nor myrrh grow in Israel, making them valuable items; they would have been carried many miles to be presented to Jesus.

How appropriate to give gifts from trees to the Messiah who was to follow in his earthly father's footsteps and learn the trade of carpentry. He must have spent many hours with Joseph, hearing about the different trees and the wood they provided, becoming familiar with the grain of each and its shaping and polishing.

It is often noted how steeped Jesus is in the natural world. His parables and teaching draw on many aspects particularly linked with the farming and fishing that would have been the main trades of many of the people with whom he talked. And of course trees feature too: in the well-known warning about not looking at the speck of sawdust in someone's eye whilst ignoring the plank in your own; in the parable of the mustard seed, and in Luke's parable of the single fig tree growing in a vineyard.[32]

As we go through Lent, we are gradually making our pilgrimage towards Holy Week, and here too – in the palm fronds of Palm Sunday, the cursing of the fig tree, the olive trees on the Mount of Olives where Jesus spent his last night, the crown of thorns, and the myrrh offered during his death – we see trees accompanying him all the way. And, of course, ultimately, we watch as Jesus is put to death on a tree: a wooden cross.

John Evelyn, the seventeenth-century English writer and gardener, expressed this beautifully:

> In a word, and to speak a bold and noble truth, trees and woods have twice saved the whole world; first by the ark, then by the cross; making full amends for the evil fruit of the tree in paradise, by that which was born on the tree in Golgotha.[33]

The Lord of all creation, the one through whom all things were made (including trees), died for us, nailed to his creation, so that we who were once far off might be brought near through his blood, spilled onto the land, and the whole creation set free. Little wonder then that the natural world responds so dramatically at the moment of his death as darkness descends over the land and the earth shakes (Matthew 27.45, 51, Mark 15.33, Luke 23.44–45).

But the story does not end there and we move from the cross to a garden, another motif that runs through the biblical story. The philosopher G. K. Chesterton, in his book, *The Everlasting Man*, wrote:

> On the third day the friends of Christ coming at daybreak to the place found the grave empty and the stone rolled away. In varying ways they realized the new wonder; but even they hardly realized that the world had died in the night. What they were looking at was the first day of a new creation, with a new heaven and a new earth; and in a semblance of the gardener God walked again in the garden, in the cool not of the evening but the dawn.

What a beautiful description of what has happened through the death and resurrection of Jesus. We are invited to join the gardener God, to walk in the garden with him knowing that Jesus the Messiah died for us and so through him, we have now received reconciliation. Jesus declares in John 15.1, 'I am the true vine and my Father is the gardener'. In the same way as we are invited to be like trees strongly planted in his life, so here too we are invited to 'abide in the vine' (v. 4): to remain closely entwined in Jesus' love so that we might bear fruit from the soil of our lives.

> Blessed be you Tree of Life,
> with your roots reaching down to the dark centre of the universe
> your leaves yearning towards the light beyond heaven.
> Shelter me with all your creation as I rise up this day
> and take my rest this night.[34]

Do trees scream?

Looking out my kitchen window I see a tree that I have watched grow for the last 25 years. A beech tree, it is now twice the size of nearby houses and beautifully shaped. In the summer, it provides welcome shade and families often sit under it while children play. I love that tree – and also the others I can see on the green, some of which I've helped plant myself – and were a developer to come and try to cut it down, I would fight hard to prevent that happening. The developer would have no qualms about felling, not knowing the tree or having any connection with it. But I do. As God saw that the plants and trees were good, so I see that tree. In some way, I have a relationship with it and I love it.

But, though I care about the trees that are around me, I am aware that I have much to learn about what goes on in their lives and how utterly amazing that is. Peter Wohlleben, who we met in Chapter Two, has written a fascinating book called *The Hidden Life of Trees: What They Feel, How They Communicate: Discoveries from a Secret World.*[35] Drawing on his experience as a forester of over thirty years and the latest scientific research, he has learnt that trees are social beings, sharing food with their own species and sometimes even with competitors. They are connected by a vast underground system of roots, interwoven with an astonishingly dense network of fungal mycelium which exchange nutrients, help neighbours in times of need, and enable trees to pass on information about insects, droughts and other dangers. They communicate with one another above ground too, as we see with the umbrella thorn acacias in the African savannah. When a giraffe starts feeding on a particular tree, that tree sends up ethylene as a warning gas to other trees in the vicinity. Immediately, they pump giraffe-repelling toxins into their leaves and the giraffes have to move some distance away to find trees that haven't had the memo.[36]

One fascinating insight is the possibility that 'when trees are really thirsty, they begin to scream' – though not in a way we can hear because it happens at ultrasonic levels. When the flow of water from the roots to the trees is disrupted, the trunk starts to vibrate. As Wohlleben says, this is probably just mechanistic, and yet:

> We know how the sounds are produced, and if we were to look through a microscope to examine how humans produce sounds, what we would see wouldn't be that different: the passage of air

> down the windpipe causes our vocal cord to vibrate. When I think about the research results . . . it seems to me that these vibrations could indeed be much more than just vibrations – they could be cries of thirst. The trees might be screaming out a dire warning to their colleagues that water levels are running low.[37]

Wohlleben has also discovered that many of the characteristics he has observed in old growth forests disappear in fields of modern agriculture, where, thanks to selective breeding, they have 'lost their ability to communicate above or below ground. Isolated by their silence, they are easy prey for insect pests'. This is also the case in planted monoculture forests, where he describes the trees as becoming more like street kids, isolated and behaving like loners.[38]

Yet trees are truly wonderful things and they give us great aestheric pleasure through the resource of their wood, their role in protecting soil, absorbing CO_2 and the way they provide habitats for a myriad of wildlife. However, deforestation continues to happen at an alarming rate, with over half of the world's tropical forests having been destroyed since the 1960s.[39]

Every region of the world has its issues with deforestation. In Africa, one of the biggest concerns is the ongoing destruction of the Congo Basin – an area that contains 20 per cent of the world's tropical forests – mostly due to clearing the land for subsistence farming.[40] In South America, the Amazon is being lost due predominantly to forest conversion for cattle ranching for beef, and just today as I am writing this, a new report has come out claiming satellite imaging has shown that an area of Amazon rainforest roughly the size of a football pitch is being cleared every minute.[41] In 2019 the sky in São Paulo turned black from unprecedented wildfires in the Amazon, many of which were from land clearances as President Bolsonaro relaxed legislation in order to open up the Amazon for agriculture and mining. In Asia the big issue is palm oil, and researchers (again using satellite imaging) found there was much greater loss than expected in the highlands of Southeast Asia.[42] In Europe, forests are faring better, with both the overall area of forest and the area of protected forest actually increasing.[43] Here, there are calls to step up foresting work and move to a re-wilding approach which focuses on the large-scale restoration of ecosystems, often re-introducing key species (such as beaver and lynx) to manage the environment in the way we saw wolves do in Chapter Two.[44] There are, however, notable exceptions even in Europe, with mass deforestation taking place in Siberia.[45]

The main driver of deforestation is agriculture, both for subsistence farming and, at a much bigger scale, for beef, soy and palm oil. Beef – including grain for cattle feed – was responsible for almost half of all forest clearance relating to agriculture between 1990 and 2008,[46] but other important crops are maize, rice, sugar cane, cocoa, tea and coffee. After agriculture, wood products (including for fuel) are the next main driver of deforestation. I remember a trip to Nigeria where I was struck by how much wood for charcoal I saw by the side of the roads. Charcoal is massive business in many countries but it is causing large-scale deforestation; that is why Tearfund's solar, biodigestor and clean cook stove projects and advocacy work are so important.[47]

The terrible impact of deforestation includes destroying biodiversity; worsening climate change; disturbing water cycles; disrupting lives and livelihoods, and human rights abuses by companies engaging in this work. A report by Global Witness states, 'It has never been a deadlier time to defend one's community, way of life, or environment', and their research has shown that agribusiness, including coffee, palm oil and banana plantations, is the industry most often linked with violence (including murder) against land and environmental defenders.[48]

As I consider the appalling consequences of deforestation, I am struck, by way of contrast, with the Jewish festival of *Tu b'Shevat*, 'the New Year for Trees'. I first encountered this when I was invited to the launch of the Eco Synagogue scheme which was held, appropriately, on *Tu b'Shevat*. On this day, the Jewish people eat a lot of fruit, particularly fruits traditionally associated with the land of Israel, sometimes also using their seeds to plant trees. I have long thought how beautiful it is to have a new year especially for trees; a day where we pause to recognize their beauty and wonder and all they do for us and the land, and to commit ourselves to looking after them and planting more.

This is not to worship trees, but rather to honour them – something that is practised in many cultures, from the Celtic 'Green Man' motif to the sacred fig trees in Kikuyuland in Kenya. Wangari Maathai, from the Kikuyu tribe herself, remembers being told by her mother that she should never collect firewood twigs from around the fig tree since it was *mũtĩ wa Ngaii*, 'a tree of God'. She also recalls that if you approached a tree during a ceremony or climbed Mt Kenya, which was heavily forested in earlier days, people had to take off their sandals, which reminds us of God's instructions to Moses as he approached the burning bush.[49] Tanzanian Andrew Kyomo talks about the proverb in his tribe that says, 'The forest is our skin and if one removes the skin of a human being, the result is death': a reminder of how important trees are to our survival.[50]

Ramine Souza and Josiani Baia are two young women who live in the Brazilian city of Barcarena in the Amazon. Having been unemployed and close to destitution, they are now jewellery makers, creating 'biojewel' adornments from the beautiful seeds around them that are plentiful in that area. They have learnt to value the relationship between the forest and the city and, alongside having a stable income, are part of a group of artisans working to protect the Amazon, raising awareness through their handicrafts, and denouncing the deforestation that would destroy not only the biodiversity of the forest, but also the businesses they have worked hard to create.[51]

They have done this through a course run by the Diocese of Amazônia of the Anglican Episcopal Church of Brazil, which has been working to educate people on looking after the rainforest for more than a decade. The course runs in the Belém region and trains leaders from riverside and remote communities, increasing awareness and changing attitudes around environmental issues, and equipping participants to develop projects to do this. The course combines protecting the environment with empowering those on the fringes of society.

Safeguarding the Amazon, which captures 25 per cent of global carbon dioxide emissions, is vital in the fight against climate change. The decline in the Amazon carbon sink in the decade to 2015 amounts to one billion tonnes of carbon dioxide – equivalent to over twice the

UK's annual emissions.[52] While vital, reforestation will not fully resolve the issue of the loss of trees because newly forested areas are not as effective carbon sinks or as biodiverse as older forests. It is therefore essential to prevent deforestation from happening, rather than simply mitigating its effects. Deforestation causes droughts, and these have led to the loss of more trees in the Amazon, in turn worsening the droughts. Environmental devastation, poverty and violence against indigenous communities are all problems being faced in the Amazon, but Bishop Marinez holds on to Jesus' words in John 10.10 that he has come to bring abundant life, and has dedicated herself and her diocese to working for 'the multiplication of a Culture of Life, the rescue of dignity, justice, peace and social and environmental preservation!'

There are schemes all around the world working to protect our vital trees, and it is encouraging when churches are involved too, as we saw with Juliana at the start of the chapter. The Church of South India, for example, has produced its Green Protocol Guidelines, approved by the Synod Executive for use in all its dioceses. It has twelve areas for action, one of which is around planting. Churches, whether urban or rural, are encouraged to plant a sapling every time there is an important function or the visit of a dignitary, and to plant fruit bearing trees in public places to provide food for other creatures. In some dioceses the first thing a couple will do after getting married is plant a sapling on the campus of the church, and guests are welcomed with the sapling of a flowering plant.[53] Green Anglicans (the Anglican Church of Southern Africa Environmental Network) is spreading a similar culture of tree planting throughout southern Africa and Kenya.[54]

In Ethiopia, the Orthodox Church 'views the natural forest as a symbol of heaven on Earth, where every creature is a gift from God and needs its habitat'. For them, these natural environments provide sites for contemplation and prayer as well as burial places, and so rural churches have instinctively looked after the forests around them whilst the rest of the country has suffered severe deforestation. Ethiopia used to be covered with tall lush forests, but this is now true for only five per cent of the country, due to mass deforestation to provide agricultural land for the huge population boom that has occurred. Aerial shots of Ethiopia dramatically show pools of green within broad swathes of brown: as forest ecologist, Alemayehu Wassie, says, 'If you see a forest in Ethiopia, you know there is very likely to be a church in the middle'. He is now working with churches to increase their understanding of the biodiversity these oases contain and how to protect them against the encroaching grazing and agricultural fields.[55] In fact, Ethiopia as a whole is taking massive steps towards reforestation. In July 2019 the nation planted an estimated 350 million trees in one day. Some think this statistic cannot have been correct but, whatever the actual figure, huge numbers of saplings were planted in a thousand sites across the country in order to begin counteracting the mass deforestation.

In the heart of London, just round the corner from Waterloo station, St John's Church has been hard at work with their churchyard – so hard at work, in fact, that it has won both a Silver Eco Church award and a Silver Gilt award for London in Bloom. Members of the congregation have planted a pollution-reducing hedge at the street-facing corner, with golden foliage that traps fumes and creates colour in a shady spot. They have dug out and planted a new wildlife garden in a disused part of the churchyard, creating a welcome sanctuary for bees and butterflies. The churchyard also features a walnut tree, two beautiful cork trees grown from seed, blossom trees, three huge plane trees, and lots of other shrubs and flowers.

Many churches all around the world, from different denominations and networks, urban and rural, are planting trees and looking after their patch of land in ways that provide habitats for other creatures and create places of beauty for people to enjoy. Lambeth 2020 will itself see a new small forest come into existence, with a tree planted in Canterbury diocese for each bishop in attendance. In this and in every action taken, the church is being what Bonhoeffer called 'a yes to God's earth'.[56]

We are called to follow God in loving and cherishing this world, not only in our understanding but in our practice too, living lives of gratitude, reverence and appreciation. Kyomo tells us, 'We cannot claim to be Christians . . . if we engage in destruction of God's creation like deforestation',[57] and yet we do so (albeit often unwittingly) when we eat a high meat and dairy diet, purchase products with palm oil in,[58] and buy wood and paper products without making sure they have come from a sustainable forest or from recycled paper. What this means is that we should all be reducing our paper usage drastically and using recycled paper wherever possible. At the very least, every Christian home and every church should use recycled toilet paper if in a country where that is available! Another really simple thing you can do is calculate your carbon emissions for a year through the Climate Stewards calculator and then offset them, directly funding tree planting and reduced-fuel cookstoves projects in Uganda, Kenya, Ghana and Mexico.[59] On top of these things, let's plant trees wherever possible as we know that this is one of the most effective and cheapest ways to tackle climate change.

Saying yes to life

On the Third Day, God created the dry land and the plants and trees and 'saw that it was good'. As we go through Lent may we also look with new appreciation at the land and trees that are around us and act in ways to love and protect them.

> To plant a tree is to say yes to life
> It is to affirm our faith in the future.
> To plant a tree is to acknowledge our debt
> to the past: seeds are not created out of nothing.
> To plant a tree is to co-operate in nature's works
> whereby all forms of life are interdependent.
> To plant a tree is to take sorrow for past mistakes;
> when we took life's gifts for granted.
> To plant a tree is to make a social statement for
> green-consciousness, for conservation and ecology.
> To plant a tree is to enhance the quality of life
> It brings beauty to the eyes and uplifts the spirit.
> To plant a tree is to make a spiritual statement
> or point. We are all members of the tree of life,
> we stand or fall together.[60]

For discussion

1 This chapter talks about a false dualism that separates out body and spirit, earth and heaven, natural and spiritual, and views body/earth/natural as inferior. Does this reflect the Christian tradition within which you stand? Where have you seen it played out? How does a fresh realization of the place of land and trees in the Bible affect your thinking?
2 How aware are you of your connectedness to the land? What helps you appreciate that more deeply?
3 In what ways has colonization impacted your country?
4 Watch this chapter's interview with Bernadette Kabonesa. She is a Senior Research Technician at the Ugandan National Agricultural Research Organization and an expert forester. You can see the interview at <www.spckpublishing.co.uk/saying-yes-resources>.
5 Consider how your actions impact on forests, whether through meat that comes from deforested land or wood and paper products you may buy. What steps could you and your church take in response?
6 Finish by reflecting on Psalm 1, asking God to root you deeply into him through the rhythms and practices of your life.

A Prayer from El Salvador

Oración por los árboles y la tierra

Dios de la creación nos has enseñado a amar la vida, a que de todos nuestros deseos debe superar el anhelo por la vida, deseo que debe trascender valorando todos los seres vivos de la creación. Los árboles y la tierra gimen a causa de nuestra poca conciencia por cuidarlos y protegerlos, nos hemos adueñado y lucrado lejos de protegerlos de la muerte. Hemos visto a la tierra como un recurso para explotar y no como madre. Aceptamos el desafío de cuidarnos a nosotros mismos para cuidar de nuestra madre y casa la tierra, de los árboles y de la vida. Reconocemos que nos has dado un entendimiento mayor al de otros seres vivos para reflejar tu carácter creativo, comunitario y amoroso con todo lo que existe. Señor, nos comprometemos a vivir cuidando de toda la naturaleza, protegiendo nuestro corazón de los deseos egoístas y viviendo no como dueños, sino como hermanos y en comunidad con todos los seres vivos y especialmente con los árboles que son fuente de vida.

Gerson Ramírez: Teólogo y miembro de la comunidad de jóvenes teólogos de Tearfund, colaborador en el movimiento Transforma Jóven de Honduras y miembro del movimiento Miqueas joven. Docente de Teología, Consultor en temas de desarrollo y teología de la misión para iglesias y organizaciones en El Salvador y Centroamérica.

Prayer for trees and earth

God of creation, you have taught us to love life. That our longing for life should be above all other desires; a transcendent longing that values all of creation's living creatures. The earth and trees groan because of our failure to care for and protect them, ruling over and profiting from them rather than protecting them from death. We have viewed the earth as a resource to be exploited rather than as our mother. We accept the challenge of taking care of ourselves in order to care for the earth our mother and our common home, for the trees and for life itself. We recognize that you have given us an understanding greater than other living creatures in order to reflect your creative, communal and loving character towards everything that exists. Lord, we commit to live caring for all nature, guarding our hearts from selfish desires and not living as proprietors, but as brothers and sisters and in community with all living things, especially the trees that are the source of life.

Gerson Ramírez is a theologian and member of Tearfund's Young Theologians, of the Transforma Jóven (Transform Youth) movement in Honduras and of the Micah Network youth movement. He is a theology teacher and a consultant on issues of development and theology of mission for churches and organizations in El Salvador and Central America.

4
Let there be lights in the sky (Genesis 1.14–19)

> [14]And God said, 'Let there be lights in the vault of the sky to separate the day from the night, and let them serve as signs to mark sacred times, and days and years, [15]and let them be lights in the vault of the sky to give light on the earth.' And it was so. [16]God made two great lights—the greater light to govern the day and the lesser light to govern the night. He also made the stars. [17]God set them in the vault of the sky to give light on the earth, [18]to govern the day and the night, and to separate light from darkness. And God saw that it was good. [19]And there was evening, and there was morning—the fourth day.

Bardsey Island is a special place, known as the island of 20,000 saints. A tiny place (just over a mile and a half long) off the north-west coast of Wales, it has been inhabited from at least the sixth century, when it was a monastic site, and holds the ruins of a thirteenth-century Augustinian monastery. In the 1800s it was a farming and fishing community, and you can now stay in the farmhouses that are dotted around the island. It is outstandingly beautiful, with views across to Ireland from one side and back to Wales and the mountains of Snowdonia from the other.[1]

One of the highlights is looking at the sky on a clear night. With no electricity on Bardsey and no nearby towns or cities to produce light pollution, the skies are truly dark, making for a majestic view. Even with the naked eye the Milky Way is beautifully clear and I remember one night lying out on the grass looking up at the multitude of stars and watching the Hubble Space Telescope and satellites going by, along with the occasional shooting star.

Take out Hubble and the satellites, and it would have been a similarly full sky that Abram gazed up at in wonder as the word of the Lord came to him and said, 'Look up at the sky and count the stars – if indeed you can count them. So shall your offspring be' (Gen. 15.5). What an amazing promise to be given to a childless man! At that point having even a single star in the sky would not have represented Abram's situation, but God tells him that his offspring shall be countless in number. The knowledge that we have now of the stars, planets, galaxies and other celestial objects in the universe is way beyond anything Abram was aware of – and we will come back to explore some of that further in this chapter – but he knew the stars in the night sky were impossible to count, and he recognized the word of God when it came to him, and so 'Abram believed in the Lord and he credited it to him as righteousness' (vv. 6).

In this chapter, our gaze turns to the sun, moon and stars made by God. We will consider the rhythm of the seasons and festivals that they were created to mark, as well as the problem of unnatural light – 'light pollution'. This chapter will feel different to other chapters as we consider eschatology (the end of the world/universe, sometimes referred to as the 'end times'), something that, in the Scriptures, is often associated with the cosmic objects created on Day Four.

Let there be lights

We have already seen God create light on Day One, but now we move into the days in which God makes the things that will populate the spaces that have been opened up – of sky, sea and land. Here on Day Four God creates the celestial objects to populate the sky and the 'spaces' of day and night.

There is a distinction between Day Four and Day One. On that First Day, God separated the light from the darkness and declared the light to be good. Hence, in Chapter One we focused on the contrast between light and dark in the Scriptures, seeing light as a symbol of God and his presence, and Jesus as the light of the world who calls us out of darkness to live in the light.

On Day Four, however, God declares both the night and the day to be good. Darkness too is important. It is a time for rest; for cool in a hot climate; for sleep. Some creatures actively thrive in the darkness, and those that do not, need the dark to sleep and recuperate in its safety. The creation poem of Psalm 104 declares:

> He made the moon to mark the seasons,
>   and the sun knows when to go down.
> You bring darkness, it becomes night,
>   and all the beasts of the forest prowl (vv. 19–20).

On Day Four, therefore, we have a different perspective on light and dark and one that declares even the dark to be good and part of God's loving creativity.

There is an important theological point being made in these verses that takes us back to our initial discussion at the start of Chapter One. We saw there that the creation narratives found their final form in the context of exile, when the Israelites were living in a strange land with a pagan religion. That religion was polytheistic, consisting of many gods who represented different aspects of the natural world. From that belief system arose the dominant creation narrative, *Enuma Elish*, which portrayed the victory of the chief god, Marduk, and the creation of the world from the body of the slain Tiamat. Having been duly praised for his victory, he assigned various jobs to the other gods to keep the created world in order. Marduk then 'fashioned heavenly stations for the great gods' Anu, Enlil and Ea, and the constellations became their astral likenesses.[2] The gods are given control of the stars and a god called Nannar is entrusted with the night and ensuring that the moon carries out its full cycle each month.[3] For Israel's ancient Near Eastern neighbours, the stars therefore represented and were ruled by deities and, as such, were thought to control human destinies. People were at the mercy of the gods in the stars.

It should be immediately obvious that this is quite different to what is being portrayed in Genesis 1.14–19. As we saw in Chapter One, there is only one God, the supreme God, who speaks the world into existence. It is noticeable that in the days we have looked at already, God names the things he has created, and yet here the author does not give the names of the sun and moon, but calls them 'the greater light' and 'the lesser light'. The ordinary names for the sun and the moon were also the names of the deities who corresponded to them, and so the author is showing that God has not created demi-gods but elements that serve a particular function. Even the line, 'He also made the stars' seems designed to counteract the prevailing narrative around the divine nature of stars. It is such a throw-away line, giving little importance to them: 'Oh yes, and he also made the stars while he was at it!'

As the *South Asia Bible Commentary* says, 'The wording makes it clear that the stars and the planets do not have power in themselves, as astrology would have us believe. They do not determine auspicious days or times. They are not to be worshipped. Like us, these planets are only creations under God's control.'[4] So our lives are not determined by the movements of the stars; rather we live our lives in communion with and within the movement of God.

Worship of the sun, moon and stars was clearly a problem for the Israelites, right from their earliest days in the desert, through the time of the monarchy (e.g. 2 Kings 21.3–5) and into the period of exile. It was such a well-known feature that even Stephen, in his speech to the Sanhedrin before being martyred, says how 'God gave them over to the worship of the sun, moon and stars' in the wilderness when they turned against him and made the golden calf (Acts 7.42–43). Such worship was punishable by death by stoning (Deut. 17.2–7), and yet even in Ezekiel's day people were 'bowing down to the sun in the east' (Ezek. 8.16).

The wider biblical material also affirms clearly that the astral objects have been created by God. Notwithstanding the rather plain description of God's creation of the stars in Genesis 1, in the psalms we are almost overwhelmed with the beautiful language used of God's creation of the heavenly bodies: 'In the heavens God has pitched a tent for the sun. It is like a bridegroom coming out of his chamber, like a champion rejoicing to run his course. It rises at one end of the heavens and makes its circuit to the other; nothing is deprived of its warmth' (Ps. 19.4–6); 'By the word of the Lord the heavens were made, their starry host by the breath of his mouth' (Ps. 33.6); 'He determines the number of the stars and calls them each by name' (Ps. 147.4), and 'The Mighty One, God, the Lord, speaks and summons the earth from the rising of the sun to where it sets' (Ps. 50.1).

In his mighty words to Job, God asks:

> Can you bind the chains of the Pleiades?
> can you loosen Orion's belt?
> Can you bring forth the constellations in their seasons
> or lead out the Bear with its cubs?
> Do you know the laws of the heavens? (Job 38.31–33)

The description of God's creation of the heavenly bodies in Genesis 1 reinforces the discussion we had in Chapter Two around creation and its relation to God. We have seen that creation comes from the space that God makes in Godself, and is therefore intimately connected to God, coming from the eternal self-giving love that circulates between Father, Son and Holy Spirit.[5] Psalm 136 picks this up, highlighting how God's act of creation is a reflection of his love: '[He] made the great lights—his love endures forever. The sun to govern the day, his love endures forever. The moon and stars to govern the night; his love endures forever' (vv. 7–9). And yet all the biblical statements around the creation of the sun, moon and stars make clear that they are separate from God as well as deeply connected to him. The creation comes from the word of God – from the eternal Son through the brooding action of the Spirit (Gen. 1.2) – but the creation is not God, and therefore is not to be worshipped.

To everything there is a season

In Chapter One we noted the symmetry and order that the author demonstrates through how he has set out his creation account, and we get this sense here too. The main reason given for the creation of the sun, moon and stars is to separate day and night and to be 'signs to mark sacred times, and days and years' (Gen. 1.14). The sense of rhythm is clear and is engrained into the patterns of the natural world and therefore into our patterns too. We live by a natural rhythm of day and night, which is encoded in our DNA, and we know how it feels when that pattern is disrupted. We shall consider Sabbath when we look at Day Seven in the Conclusion, but that is another pattern given to us: a rhythm of resting every seven days. As human beings, we have naturally sought to understand time and give it shape. We have divided time into segments that we can comprehend, some of which are determined by physical factors (such as the time it takes for the moon to rotate around the earth), some of which we have devised (such as dividing hours and minutes into segments of sixty). Every culture has their way of capturing and expressing time. For example, the current epoch of the Islamic calendar, the Hijri era, began in 622AD, the year that Muhammad and his followers migrated from Mecca to set up the first Muslim community at Yathrib (now Medina). Each year is 354 or 355 days, making this year 1441, according to Muslim calendars, rather than 2020.

The seasons provide a basic rhythm for our lives, whether through the pattern of the monsoon and dry periods; the long and the shorter rains, or the framing of spring, summer, autumn and winter. We spend a lot of energy either trying to enjoy and get the most out of those seasons or protecting ourselves against them.

After the flood, God promises, 'As long as the earth endures, seedtime and harvest, cold and heat, summer and winter, day and night, will never cease' (Gen. 8.22). Following what we saw in the previous chapter about the significance of land, it is no surprise that the regularity of the seasons is of such importance to the Israelite people. It also functions as a metaphor. The Lord says through Jeremiah, 'Even the stork in the sky knows her appointed seasons, and the dove, the swift and the thrush observe the time of their migration. But my people do not know the requirements of the Lord' (8.7). Knowing the requirements of the Lord should be as built-in to his people as the natural seasonal callings are to the birds.

Many of us reading this now live lives that bear little relation to the seasons and we can go through the years disconnected from the changing world around us. Living in the UK, I am forever grateful for central heating and I am not advocating being so in tune with the seasons that I wake up in the winter with frost on the inside of the windows! But, finding ways to be alert to the patterns of the natural world as they change through the year can help link us to our own patterns and seasons. One of the habits I have developed working at the Tearfund office in southwest London is to take a 45-minute walk at the end of the day through the nearby park. The second largest royal park in London, it is beautiful with many mature trees and deer (and now flocks of green parakeet too). One of the pleasures of this practice is seeing the park and its inhabitants grow and change as the months go by – a regular cycle of the seasons, of death and new life – and I walk with God there in the evenings as a reflection of my whole life's walk with God, and the seasons of death and life that brings too.

This finds resonance with the words of the Teacher in the book of Ecclesiastes: 'There is a time for everything, and a season for every activity under the heavens' (Ecc. 3.1). The Teacher's words are set in the midst of his reflection on the futility of life and the pointlessness of doing anything more than enjoying food and drink. Underlying the gloomy reality though is the theme of wisdom and the benefit of seeking it rather than riches. One of the characteristics of the wise is the simple acknowledgement that there is a time for everything. Life can be long. Circumstances change. What is now is not what will be in the future. We cannot hold on; we must let go and allow life to unfold. So the recognition that all things have their seasons can bring comfort and stability to our lives though they often feel anything but stable.

The Celtic Christians are well known for being a people strongly rooted in the natural seasons, and many aspects of their faith arose from that. The *Carmena Gadelica* is a collection of pre-Christian and Christian prayers, blessings, invocations and other folklore gathered by Alexander Carmichael in the nineteenth century, reflecting the oral traditions of the Scottish highlands.[6] There are prayers and blessings for many aspects of daily life, for different elements of the natural world, and for the rhythms of day and night, waking and sleeping. The sun and the moon are a part of these traditions, and of course the classic Celtic cross has the round circle of the sun in the middle, 'held in the arms of the cross'. As Newell says, this was not nature worship, this was 'Christ-mysticism that reverenced nature'.[7] Let us hear one of these sun prayers:

The eye of the great God,
The eye of the God of glory,

The eye of the King of hosts,
The eye of the King of the living,
Pouring upon us
At each time and season,
Pouring upon us
Gently and generously,
Glory to thee
Thou glorious sun.
Glory to thee, thou sun,
Face of the God of life.[8]

Tess Ward, in the prayer book we have encountered in previous chapters, bases her prayers on the themes that arise through the seasonal and festive rhythms of each year. They are particularly rooted in the traditional solar festivals of the northern hemisphere: the winter solstice when the day is at its shortest (December 20–23); the summer solstice when the day is at its longest (June 20–23), and then the spring equinox and autumn equinox (March and September 20–23), when night and day are equal. Solstice is simply the Latin for 'the standing of the sun'. Between those four points are the quarter days, which are the lunar festivals celebrated at the full moon. Imbolc is the first, in February; Beltane is at the beginning of May; Lammas is at the start of August, and the Samhaim is at the beginning of November.[9]

Traditionally, these solar and lunar festivals have been associated with paganism and the Church has steered away from them. But, a renewed appreciation for God's creation of the sun and the moon can give those of us who live in seasonal parts of the world confidence that we can use these points that mark the turning of the earth towards and away from the sun. They can be times for us to reflect on the passing of time in our lives, and on themes of darkness and light and gratitude for God's continued involvement in our world – even in those places and situations that seem darkest. And we can use them to consider how the seasons are changing because of the climate crisis, and as a prompt to commit ourselves to action.

Overlaying the rhythms of day and night, the natural markers of time, and the seasons of each year (however those look to us, where we live in the world), there is also the rhythm of the Church, which has its own seasons and festivals. Readers of this book will come from a host of different denominations and networks: you may relate to a tradition that does not use any sort of church calendar beyond Easter and Christmas, or you may follow a full seasonal cycle of Advent, Christmas, Epiphany, Lent, Holy Week, Easter, Pentecost and Trinity – with Ordinary Time in between! Whichever kind of church we are a part of, there is benefit in discovering the richness of the annual rhythm of the Christian faith, as it helps us remember and dwell on the saving work of Christ and often links our faith with the agricultural year. (Churches in some countries, for example, will celebrate Rogationtide, Lammastide and Harvest.) This both honours those in our churches who are working on the land and helps those of us who are not to appreciate our dependency on it.

One 'sacred time' that is growing in popularity is the Season of Creation (sometimes called Creationtide), which runs from September 1 to October 4, the feast day of St Francis. This is a period when churches around the world can particularly focus on caring for God's wider creation, through their worship, teaching and practical action. There are good resources available online to help with this.[10] But, whether or not you want to make use of a particular set time, it is good for us to think about how we celebrate the sacred moments of our church lives, and whether the understanding of creation and God as creator that we have been developing through this book is part of those times, or whether it is neglected. As British Baptist minister and theologian Chris Voke says, 'The vision of the triune God as the creator, recognition of the createdness and dependence of human beings upon God, and a stress on responsibility towards the world in which the Creator has placed us . . . are essential parts of the story [that should be] told in Christian public worship.'[11]

One way to do this is to get outside! You could go out as a church to a nearby park or into your churchyard and hold a service there, closer to the natural world. You could collect leaves and reflect on the verses we looked at in Chapter Three from Psalm 1, thinking through how we might meditate on God's word and be planted in his stream of living water. You could do a litter pick as a church, taking time together to pray for the people whose rubbish you have collected and asking God to speak to you about their problems and struggles. Or you could spend time looking at the sky and reflecting on Psalm 19: 'The heavens declare the glory of the Lord'. Ask yourself what strikes you about the sky today and look out for an image in the clouds that reminds you of something about God.[12] Though it is important to be mindful of those with limited physical abilities, the possibilities are endless and you will be surprised at the new life breathed into you and your church as you worship God in the midst of his creation.

Gazing into space

Throughout *Saying Yes to Life*, we are focusing on the world in which we live; the God who made it; our place within it, and how we are to live in relation to the rest of what God has made, both human and non-human. In this chapter, however, Day Four causes us to look beyond this world into the incredible, dizzying space that is . . . well, space, and to remember *all* of this was made by God.

The world is one tiny piece within a vast universe – so vast that I, at least, can scarce comprehend it. The world we inhabit is one planet within a solar system . . . within a galaxy . . . within the universe. Our sun is just one of between 200 billion and 400 billion stars in the Milky Way galaxy, and earth is just one of at least 100 billion planets. There may also be ten billion white dwarfs, a billion neutron stars and a hundred million black holes. And that is just one galaxy out of possibly two trillion galaxies![13]

The moon is 240,000 miles from the earth, which is the average distance walked by a human being in their lifetime, and if you imagine the sun to be the size of a peanut then the earth would be a grain of salt on its surface. To represent the distance to the nearest star, another peanut would need to be taken 200 miles away. However, that is just a trip round the corner compared to the furthest object seen in our universe, which is GN-z11, a small galaxy observed by the Hubble Space Telescope. It is so far away that the light we observe set off from the galaxy 13.4 billion years ago.

As human beings we are constantly fascinated by whether life might exist beyond our world, as witnessed by all sorts of films and TV series and by the number of claimed UFO sightings. More than three and a half billion years ago, Mars did indeed have the potential for life. It was a blue planet with a lot of its surface covered in water, and organic compounds found in sedimentary rocks have indicated that some of the building blocks of life were present. We may yet discover evidence of the simplest lifeforms on Mars, or indeed on other planets and in places still to be discovered. Mercury, Venus, Earth and Mars all have had the potential for life at some stage in their history, but only Earth has become the flourishing planet that it is today.[14]

In our previous chapter we looked down at the ground beneath our feet and what has sprung forth from it. Now, we look up into the vast magnitude of space and see there a different kind of beauty and wonder. Saturn's rings are asteroids, moons and comets, shattered into billions of tiny frozen ice chunks by Saturn's gravity and pulled into gravitational orbit around the planet, glittering brightly. Gas giant Jupiter is a swirl of cold, windy clouds of ammonia and water, floating in an atmosphere of hydrogen and helium, and its famous Great Red Spot is a giant storm bigger than Earth that has been raging for hundreds of years. Close-up images reveal vivid colours twirling into stunning patterns, looking like an abstract painting of oils and watercolours. By contrast, Uranus exists in cold, white, icy magnificence.

Our understanding of the universe is constantly evolving. In Chapter Three we saw how Solomon was known for his great wisdom and insight about plants and trees. How he would have marvelled at the work being done by astronomical scientists today as they push back the frontiers of our knowledge about space! And how we wonder too at the vastness of the universe and at the fact that, amidst the billions of galaxies, stars and planets, we are here, on a tiny planet that is teeming with life. We can only join in with the Psalmist in exclaiming,

> When I consider your heavens, the work of your fingers,
> the moon and the stars, which you have set in place,
> what is mankind that you are mindful of them, human beings
> that you care for them? (Ps. 8.3–4)

It is not only we human beings who praise the Creator: the heavenly bodies join in too:

> The heavens declare the glory of God;
> the skies proclaim the work of his hands.
> Day after day they pour forth speech;
> night after night they reveal knowledge.
> They have no speech, they use no words;
> no sound is heard from them.
> Yet their voice goes out into all the earth,
> their words to the ends of the world (Ps. 19.1–6).

In Psalm 148, the Psalmist calls on them to join in with the rest of creation in praising our Creator:

> Praise the Lord from the heavens; praise him in the heights above.
> Praise him, all his angels; praise him, all his heavenly hosts.
> Praise him, sun and moon; praise him, all you shining stars.
> Praise him, you highest heavens and you waters above the skies.
> (Vv. 1–4)

Polluting the night skies

The experience on Bardsey Island with which we started this chapter was so memorable for me because it was so rare. Most nights I consider it a clear night if I can see Orion and the Big Dipper, and it is a sad reality that most of us are seldom in places that are dark enough at night for us to enjoy the stars in all their splendour. In fact, light pollution is now so bad that more than one third of the human population is no longer able to see the Milky Way.[15] In Chapter One we reflected on NASA's 'black marble' images, realizing that the earth at night is electric with lights criss-crossing the globe.

Light pollution is a challenge not only because it prevents us enjoying the aesthetics of a starry night sky, but because it can have fatal consequences for creatures that use both the dark and light in particular ways. One significant marine problem is the death of thousands of baby turtles each year in Florida. When they hatch on the beach (generally at night but also in the early morning or late afternoon) they have an inbuilt orientation away from the dark shapes of the dunes towards the brightest direction, which naturally is the sea reflecting the night sky. However, the Florida beaches are lined with properties with artificial lighting that draws the hatchlings away from the sea, ultimately leading to their death by dehydration, predation or being run over.[16]

Another aquatic trial is the impact of artificial lighting on frogs and toads. Our world's amphibians are in a state of terrible decline: 40 per cent of all amphibians are threatened with extinction. This is due to a range of factors, such as disease, habitat destruction and their susceptibility to toxic chemicals in waterways, but research on the American toad showed that toads born in places with artificial lighting changed their natural behaviour and did not grow as well as those born in places with limited or no artificial light.[17]

Problems associated with light pollution do not only affect water creatures; numerous studies show that artificial light at night causes disruption for millions of migratory birds. It can lead them to become disoriented, and can change their flight behaviour, their sleep patterns and their reproductive capacity. One particular factor has been highlighted with regard to birds that fly at night and use flight calls to communicate with their flock and navigate their passage. Researchers have found that levels of flight calling intensifies when birds are over illuminated areas, suggesting they are becoming increasingly disorientated and need to communicate more. Moreover, although it is not fully understood why, it was also found that flight-calling birds have higher rates of death by crashing into illuminated buildings.[18]

To help us appreciate the wonders of the night sky and to enable other creatures to flourish, we need to reduce the amount of light we produce at night. For those of us in countries where solar lamps in the garden are fashionable, we can decide not to use them, however pretty they might look. In many countries, where shops and offices leave their lights on, we could ask them either to turn them off or, if security is an issue, reduce the brightness. And likewise, we can contact our local authorities and ask them to turn off or lower street lighting and install downward facing lights that transmit less light upwards into the sky. If we are part of a church that shares its building with bats, we can help by ensuring we turn off lighting that is near a bat roost, so as not to cause disturbance and potential abandonment.[19]

Hands that flung stars into space

Graham Kendrick, in his beautiful worship song, 'From Heaven You Came (The Servant King)', wrote of how 'hands that flung stars into space' are also hands that 'to cruel nails surrendered'. This provides a glorious way to continue the link between creation and redemption that we have seen in previous chapters, as we are reminded that the Saviour Jesus Christ – the one who in the closing words of the Bible is called 'the bright Morning Star' (Rev. 22.16) – is also the Creator Son of God.

At various points throughout this book we have noted how the work of Jesus is rooted in the natural world – not only in his parables and teaching but also through the seminal moments of his life. He is the Lord of all creation and it should therefore be no surprise that this is the case, or that the winds and the waves obey him. Maybe what *is* surprising is how often we do not notice it!

Perhaps most obviously in relation to this chapter's focus, Jesus' birth is announced by a great heavenly company of angels, who appear to shepherds on a hillside with their sheep, and by a star that alerts the magi and leads them to where Jesus lies. The Word who was with God and was God in the beginning; the Word through whom all things were made; the Word who brings light into the world . . . his birth is accompanied by a night sky illuminated by the glory of the Lord and by a bright shining star. The heavens do indeed declare the glory of the Lord as, in the incarnation, the Creator takes on the flesh of his creation, and comes to live among us so that we might be redeemed and brought back to life in God. And then, as we have noted previously, his death – as he bears our sin in his earthly body – is accompanied by a dramatic response in the natural world, as the sun stops shining.

British Bible scholar, Richard Bauckham, has written a beautiful poem called *Song of the Shepherds* in which he describes the angelic visitation. Here is one extract, written from the shepherds' perspective:

> They say that once, almost before time,
> the stars with shining voices
> serenaded
> the new born world.
> The night could not contain their boundless praise.
>
> We thought that just a poem –
> until the night
> a song of solar glory,
> unutterable, unearthly,
> eclipsed the luminaries of the night,
> as though the world were exorcised of dark
> and, coming to itself, began again.

Bauckham reflects passages we have already looked at (Job 38.4, 7; Ps. 19.1–2) and shows the shepherds in a moment of illumination, seeing the truth of those scriptures as the Light of the World is born and the world comes to itself and begins again.[20]

Sun, moon, stars and the end of the world?

The Bible is full of references to the heavenly bodies created on Day Four, reminding us that the story of salvation unfolds within the physical world: a world of day and night; sunrises and sunsets; starry nights, and the ever continuing rhythm of the seasons. Through it all is the recognition that these daily and yearly patterns come from God and are part of how he has ordained the world to be: 'The day is yours, and yours also the night; you established the sun and the moon. It was you who set all the boundaries of the earth; you who made both summer and winter' (Ps. 74.16–17); and it is God who 'causes his sun to rise on the evil and the good, and sends rain on the righteous and the unrighteous' (Matt. 5.45). The astronomical bodies both praise God themselves through the very wonder of their existence, and inspire our praise of him too, as we marvel at the enormity of what lies beyond us. But there is one particular way in which the sun, moon and stars are used in the Bible that we need to spend some time considering, and that is their use in texts that take us into the area of eschatology, or talk about the 'end times' (coming from the words *eschatos*: 'last', and *logos*: 'word').

When I was a young pre-teenager, I loved a Christian band called The Reps. One of my favourite songs on my cassette tape had the chorus, 'And the Moon shines red in the sky, And we can't change a thing, however we try. The trumpet sounds and the people run, but they cannot hide, Jesus Christ has come'. Some of those words are from Scripture, with the trumpet sound a reference from 1 Cor. 15.52. In particular, the moon shining red comes from Joel's prophecy about 'the last days' when God will pour out his Spirit on all people, which Peter also uses in his speech at Pentecost. God says:

> I will show wonders in the heavens above
> and signs on the earth below,
> blood and fire and billows of smoke.
> The sun will be turned to darkness
> and the moon to blood
> before the coming of the great and glorious day of the Lord.
> (Acts 2.20–21)

This is just one of a number of references in the Bible, particularly from the time of the Old Testament prophets and into the New Testament, in which the sun, moon and/or stars are used to talk about the future times (eg. Isa. 13.9–10; Isa. 34.4; Lk. 21.25; Rev. 8.10–12).

What do these passages mean? Is the language to be taken literally, along with other passages that seem to speak about what will happen at 'the end' or are there other ways of understanding it? Certainly there is a very strong strand within Christian theology that does view such passages as plain fact. The Left Behind series is one of the most obvious examples of this:

> The Bible teaches that at a time no one knows (thus all date-setting is folly), Jesus Christ will appear from Heaven in the clouds and true believers will be caught up to be with Him (I Thess. 4.13–18), then comes the Tribulation (Rev. 6–19), culminating with the Battle of Armageddon and the Glorious Appearing (Matt. 24.29–31 and Rev. 19.11–21). Finally, the Millennium (the 1,000-year kingdom of peace when Christ shall rule while Satan is bound), after which He will set up a new heaven and new earth (Rev. 21).[21]

However, in order to understand these biblical passages correctly, we need to see that they fit within the genre of apocalyptic writing. Apocalyptic literally means 'revelation' and there is a strong tradition within Jewish literature of a heavenly being giving someone a vision of the future or of another dimension of worldly reality. The aim of this type of writing is not so much to answer questions we may have about how the universe is going to end and what events might take place as that happens. Rather, it is to speak to the people of God about their current situation – be that exile or persecution – to lift their eyes above their hardships in order to gain the 'God's-eye view' on where they are and an understanding of their place in the unfolding of history. As Richard Bauckham explains in his writing on the book of Revelation, 'The effect of John's visions . . . is to expand his readers' world, both spatially (into heaven) and temporally (into the eschatological future) or, to put it another way, to open their world to divine transcendence . . . It is not that the here-and-now are left behind in an escape into heaven or the eschatological future, but that the here-and-now look quite different when they are opened to transcendence.'[22]

The language of apocalyptic writing is not meant to be taken literally. If I were to tell a friend 'my world has collapsed', they would know that something terrible has happened, but would not assume that the physical world around me had literally fallen apart. Or, to use an English expression, if I were to run indoors soaking wet, and exclaim, 'It's raining cats and dogs out there!', you would expect it to be pouring with rain – you wouldn't expect to see cats and dogs literally falling from the skies. Apocalyptic language is poetic and evocative, designed to express theological realities and the enormity of events such as the overthrow and downfall of an oppressive regime. With that in mind, we need to be careful not to interpret words that are about an event in this present time and age with words that are concerned with the final future of the world. Sometimes it can be hard to distinguish between them. For example, the passage from Luke about the coming of the Son of Man (21.25–28) has traditionally been taken to describe the signs that will accompany Jesus' second coming. Tom Wright, however, argues strongly that this is a misinterpretation and that Jesus' words are actually about the forthcoming fate of Jesus on the cross and his resurrection, and the oncoming fall of Jerusalem that took place in AD70.[23] This is not to deny that there are other places in the New Testament that talk about Jesus returning to his world (1 Thess. 4.13–5.11 being the obvious one), just that this passage, with its references to wars, earthquakes, pestilences and famines, should not be taken out of its original, historical context and made into something that it is not.

The conclusion we can draw, therefore, is that passages about the moon turning blood red and the stars falling from the sky shouldn't be taken literally, nor are they always about the 'end times'. By way of another example, Joel's words about the day of the Lord in 3.15 – 'the sun and moon will be darkened, and the stars no longer shine' – are clearly about the immediate, historical restoration of Israel to her land after exile, not about an end-of-the-world 'day of the Lord'. We must be wary of lifting passages out of the Scriptures and claiming they tell us exactly what is going to happen at some indeterminate time.

However, that does not mean we can say nothing about the future of creation. Indeed, it is vitally important that we understand what we can say because what we believe about this impacts how we live today. It is worth noting that when questions of eschatology feature in the New Testament letters, it is always to inspire a response in the present: 'Therefore encourage one another and build each other up, just as in fact you are doing' (after Paul's words about the coming of Christ, 1 Thess. 5.11); 'Therefore, my brothers and sisters, stand firm. Let nothing move you. Always give yourselves fully to the work of the

Lord, because you know that your labour in the Lord is not in vain' (after Paul's discussion on the resurrection body, 1 Cor. 15.58); 'What kind of people ought you to be? You ought to live holy and godly lives as you look to the day of God and speed its coming (after Peter's teaching on the eschatological day of the Lord, 2 Pet. 3.11–12). As German theologian Jürgen Moltmann has said, 'From first to last, and not merely in the epilogue, Christianity is eschatology, is hope, forward looking and forward moving, and therefore also revolutionizing and transforming the present'.[24]

So how does eschatology impact on the things we are looking at in *Saying Yes to Life*? If what we believe about tomorrow affects how we live today, that is no more true than when thinking about how we live as Christians in relation to the wider creation around us. At every talk I give and every interview I do for the Christian media, I can guarantee I will be asked a question along the lines of, 'But what about the belief that this world is going to be destroyed? Does that mean it's a waste of time to care for it now?' There are a number of excellent books on eschatology that I recommend reading if this is something you would like to think through more fully. I cannot hope to do justice to the topic here or cover every relevant part of the Bible, so please do look at the 'further resources' section at <www.spckpublishing.co.uk/saying-yes-resources>. Nonetheless, there are some themes that are important for us to consider in the context of this chapter.

I can remember as a little girl standing outside my parents' back door on the patio, looking up to the skies and saying to Jesus, 'Dear Jesus, please could you say hello to Granny from me and give her my love?' I had no carefully considered theology on which to base my prayer, but a natural instinct that, wherever Granny was, she was with Jesus in some way. And actually I believe that aspects of that prayer are still right and that Granny was – and is – being held in Jesus' presence until the time when he returns (though whether or not Jesus can pass on a message from me to her is a question that will have to remain unanswered for now).

I can also remember thinking about the concept of eternity and being with God in some immaterial, ethereal place for ever . . . and ever . . . and ever . . . and ever . . . and experiencing something like vertigo at the thought, feeling a bit guilty because I wasn't sure if it really sounded that appealing! Thankfully, my more recent consideration of the subject has given me a different perspective, one that I believe is more strongly rooted in Scripture.

For many years, the dominant view within Christianity has been that, at some point in the future, God will destroy this world in judgement and we will spend eternity in heaven. In popular culture and classical art this has been portrayed as a place in the skies with clouds and angels strumming harps. Billy Graham was very clear when he said, 'My home is in heaven. I'm just traveling through this world', and in the 1970s, Christian rock legend Larry Norman reflected a similar view with his album, 'Only Visiting This Planet', and the song, 'I Wish We'd All Been Ready'. This was about the view held by some Christians that when Jesus returns, those who are still alive will be caught up in the air and taken off to heaven, along with those already dead who will be raised, leaving behind those who are not Christians (an interpretation known as the Rapture). Wangari Maathai talked of how a popular song at funerals in Kenya is the American Gospel one, 'This World is Not My Home', with the lyric, 'My treasures are laid up somewhere beyond the blue'.[25] We see this view today when we sing the wonderful hymn, 'How Great Thou Art', based on a nineteenth-century Swedish poem, with the line, 'When Christ shall come with shout of acclamation, and take me home . . .'.

So what and where is our home? Is it true that it is not the earth and we are 'only visiting this planet'? And, wherever our home is, what is the future for this world and indeed for the universe? In order to work towards some answers, let us start by reminding ourselves of a number of key things we have explored over the course of this book so far. In Chapter One we saw the link between redemption and creation: that the saving God is also the creating God. As Paul affirms in his opening declaration to the Colossian church (Col. 1.19–20), Jesus' blood shed on the cross has brought peace for all things – not only people, but all things in heaven and on earth. In Chapter Two we considered the concept of God withdrawing to make space for his creation, and therefore the inherent connection between God and that which he has made. In Chapter Three we reaffirmed the essential goodness of creation and the need to avoid a dualism that creates a false separation between physical and unphysical, earth and heaven, and sees earth as inferior to heaven. God's physical creation is loved by him and given value because he looks at it and declares each part of it good. We also saw the importance of the land as an integral part of the story of salvation; that it mattered to God how his people treated the land and its inhabitants – both human and non-human – and that the glimpses of the future in Revelation 21 and 22 give a picture of a garden city with land, trees and water.

The biblical passages that relate specifically to eschatology are tricky, particularly in the New Testament, with strange images that can be difficult to interpret. However, the themes above form the foundation for how we navigate biblical eschatology as a whole. So let us look now at three key passages that are important to this understanding.[26]

The first is Isaiah 65.17–25. There are many other passages in the Old Testament where prophets speak of the future hope of God's people, but this one needs noting because it is the passage that the prophet John quotes when he talks about seeing a 'new heaven and a new earth' (Rev. 21.1). The context for these words is the time when the people had returned from exile and were looking forward to the rebuilding of their nation (the same sort of time period covered in the books of Ezra and Nehemiah).[27] The words are clearly spoken into that situation, but contained within them are the seeds of a wider hope for the eschatological future. We see here what we have noted already: the future hope that developed in the Old Testament, upon which the earliest Christian thinking was built and with which Jesus would have been so familiar, had a very physical dimension and encompassed the wider natural world and human society, people and animals living together peaceably.

Given the historical context of these words, it becomes immediately obvious that when this passage talks about 'a new heavens and a new earth' it does not literally mean the creation of completely new entities, but is speaking poetically to describe a situation of radical renewal. My elder daughter recently took a pair of scissors and a needle and thread to a t-shirt of hers, did some clever things with them, came downstairs proudly and said to me, 'Look Mum, I've got a new top!' It was clearly the original t-shirt, but it looked very different: it had been transformed.

The second passage is 2 Peter 3, in which Peter writes about the day of the Lord and what will happen. Verse 10 in particular has been a foundational text for the view that the world will be destroyed, based on the King James Version translation which reads, 'But the day of the Lord will come as a thief in the night; in the which the heavens shall pass away with a great noise, and the elements shall melt with fervent heat, the earth also and the works that are therein shall be burned up'. From that reading a destruction-of-the-earth view is entirely understandable. However, as Steven Bouma-Prediger has put it, 'This verse represents perhaps the most egregious mistranslation of the New Testament'.[28]

Our Bible translations are based on a variety of different manuscripts. There is not only one New Testament in the original Greek which translators then translate into their own language: there may be a number of manuscripts for one passage, dating from different times. These manuscripts often have variations in them and scholars have to decide which ones are the most reliable, one of the accepted criteria being that earlier manuscripts are best. In the case of 2 Peter 3.10, the KJV is based on a couple of manuscripts that use the Greek word, *katakaesetai*, 'will be burnt up' for the end of the verse, but most of the manuscripts considered the earliest and therefore the most reliable, actually use a different Greek word, *heurethesetai*. New Testament scholars think that those who wrote those manuscripts did not understand why *heurethesetai* was used and therefore changed the wording so it made more sense to them.

Heurethesetai carries the meaning of 'will be found' or 'will be discovered' (similar to 'eureka!') and is a positive term. It links to the words earlier in the chapter about creation and the flood (where the flood brought about a new creation that had not entirely obliterated the first, but had destroyed that which was sin and not of God). It also links to the fire of judgment of verses 7, 10 and 12 which, reflecting Malachi 3.2–3, can be seen as akin to the refiner's fire which burns up the dross so that the good might shine through.[29] The South African

Revd Rachel Mash, who we met in Chapter Two, told me that after recent fires in Cape Town, an amazing array of flowers bloomed, which gave her a fresh perspective on this passage. Tom Wright acknowledges that 2 Peter 3.10 is 'a difficult and obscure text, and likely to remain so', but reaches this conclusion:

> The worldview we find is not that of the dualist who hopes for creation to be abolished, but of one who, while continuing to believe in the goodness of creation, sees that the only way to the fulfilment of the creator's longing for a justice and goodness which will replace the present evil is for a process of fire, not simply to consume, but also to purge.[30]

Our third passage is Revelation 21 and 22, with its beautiful picture of 'a new heaven and a new earth'. It describes the Holy City, Jerusalem, coming out of heaven, with the river of the water of life flowing through it from the throne of God, and the trees of life (that we first meet in Genesis 2.9) standing on each side of the river, bearing fruit and giving their leaves for the healing of the nations. It is a truly stunning conclusion to the biblical story, and one we have commented on a few times already as we have seen how our themes of light, water, land and trees have found their place in this final narrative.

There is a great deal that may be said about these chapters, but we will restrict ourselves to two points. First, we need to note the recurrence of the new heaven/new earth language we have already encountered in both Isaiah 65 and 2 Peter 3. In New Testament Greek there are two words for new – *kaine* and *neos* – and they would appear to be used interchangeably. In the English language, we tend to think of 'new' as meaning something completely different – if I got a new car it would be a totally different entity to my old car. But, in a similar manner to my elder daughter's t-shirt, both Greek words can sometimes mean something that is not totally new, but rather something that is renewed or transformed.[31]

So, for example, Paul talks of there being a 'new creation' if anyone is in Christ (2 Cor. 5.17). When you became a Christian you didn't become a totally new human being, one your friends could not recognize and had to ask, 'Who are you?' And yet . . . maybe you did become someone so transformed that your new behaviour prompted them to wonder, 'Who are you?!' The Greek for 'New Testament' is *kaine diatheke*, literally 'new covenant'. We do not understand this to mean that the Old Testament is replaced and no longer needed (so we stop reading it, as happened with the Marcionite heresy), but as being fulfilled and transformed.

As we have considered already, such a use of 'new' carries within it both continuity and discontinuity. We get some insight into this when we think of Jesus' post-resurrection body. There was continuity in that he could walk, break bread and cook, and Thomas could physically touch his scars; and yet there was discontinuity as he could appear and disappear and he was not always immediately recognizable. One of our key challenges as we envision what a renewed heaven and earth might look like is to hold together both continuity and discontinuity, not allowing either to overpower the other. Too much discontinuity negates the biblical witness that this physical universe is valued by God and will not be completely replaced; too much continuity negates the witness of contemporary science that this universe will come to an end, and the need, therefore, for the transforming action of our faithful God.[32]

Second, Tom Wright (and others) has long argued against the view that our eternal future is in a non-physical heaven: heaven ('God's space') is a place where we will rest while we wait for the Lord to return.[33] Our final destiny is on the united and transformed heaven and earth, within a transformed universe.[34] It seems to me that the picture we are given in Revelation 21 and 22 is a wonderful vision of just that reality. The Holy City comes down out of heaven to earth, but it is a transformed earth unlike the one we know now. There is no more suffering or death, no more sadness and tears (quoting Isaiah 25.8). Human relationships are restored, and God now dwells with his people. The sea – representing chaos and wickedness, as we saw in Chapter One – has gone, symbolizing that all evil has been destroyed. The vision of heaven in Revelation 4, with its description of the living creatures praising God, presumably does not change when heaven comes down to earth, leaving us finally with a holistic picture of God and his creation, living and worshipping him together as we were originally intended to do.

Living in hope

Where does all this talk about heaven and earth, sun, moon and stars leave us as we draw this chapter to a close? Perhaps the first thing to say is that our call to take care of this world and all its creatures – including human creatures of course – does not ultimately rest on any particular eschatological view. As we are seeing throughout *Saying Yes to Life*, there is plenty else in the Scriptures that leads us in that direction.

Nonetheless, for too many people the belief that this world is going to be destroyed has been held hand-in-hand with the assumption that we need not bother looking after it now (a view that does not rest on strong biblical foundations).

What we have explored instead is a theology of the future that anticipates God transforming this present reality. Romans 8 looks forward to this time, when the children of God will be revealed and creation will be set free from its bondage to decay. As Tom Wright says, 'Creation will enjoy the freedom which comes when God's children are glorified – in other words, the liberation which will result from the sovereign rule, under the overlordship of Jesus the Messiah, of all those who are given new, resurrection life by the Spirit'.[35]

This is a wonderful vision that motivates our hopeful action today. Director of Langham Partnership, Chris Wright puts it like this:

> Ecological action now is both a creational responsibility from the Bible's beginning, and also an eschatological sign of the Bible's *ending* – and new beginning. Christian ecological action points towards and anticipates the restoration of our proper status and function in creation. It is to behave as we were originally created to, and as we shall one day be fully redeemed for.[36]

Therefore, every action we choose to take that looks after this world (even when it's raining . . . or too hot . . . or inconvenient . . . or maybe more expensive . . . or not the usual thing to do and makes us look different . . .) shows our wish to live in anticipation of the future that Jesus' death on the cross and the presence of his Holy Spirit guarantees, and we move towards towards the future glory that God will reveal through his creation.

For discussion

1 Don't miss this chapter's interview with Professor Sir Martin Rees, Astronomer Royal and former Master of Trinity College, Cambridge, and President of the Royal Society. You can watch it at <www.spckpublishing.co.uk/saying-yes-resources>.
2 How much is the wider creation brought into your church worship gatherings? Is it naturally incorporated into your prayers, songs, liturgies (if you use them) and sermons? How could you help that happen more? Use the suggestions earlier in the chapter or the online resources to do something outside as a church.
3 This chapter has had a strong focus on eschatology. Are there things in it that are new to you or that you don't understand? If you are in a Lent group, talk these things through together.
4 Do you feel the stretch of living in 'the overlapping of the ages'? How do you manage that tension? In what ways does the eschatology in this chapter motivate how you live now?
5 Look back over the last chapters and reflect on – or discuss with others – what things you have committed to do and whether or not you have done them.

A prayer on the sun, moon and stars from the Philippines

Our heavenly Father, as we look up to you in the vastness of the skies,
The sun that you have made opens our eyes to a world lit in colour and clarity,
And the moon and the stars remind us of your faithfulness and steadfast presence,
Amidst the seasons of darkness and our community's moments of uncertainty.
Lord Jesus, you have shown us how from beginning to end was the light of love,
That as endless as the heavens above so is the grace that sustains all things,

So with faith that the Spirit has wrought in us, we seek the care every creature is to have,
As we dream, hope, and labor for a future wrapped in the fullness of joy that your new creation brings.

Amen.

Rei Lemuel Crizaldo is an artist, a local author, and advocate of doing integral mission based in Manila, Philippines.

5
Let the waters teem with living creatures and let birds fly (Genesis 1.20–23)

> [20]And God said, 'Let the water teem with living creatures, and let birds fly above the earth across the vault of the sky.' [21]So God created the great creatures of the sea and every living thing with which the water teems and that moves about in it, according to their kinds, and every winged bird according to its kind. And God saw that it was good. [22]God blessed them and said, 'Be fruitful and increase in number and fill the water in the seas, and let the birds increase on the earth.' [23]And there was evening, and there was morning—the fifth day.

A few years ago I attended the A Rocha forum in Portugal. Every three years, leaders from the various national A Rocha organizations around the world get together to meet one another, talk, cry, pray, inspire, encourage, worship, strategize, laugh and eat together.[1] People came from 21 different countries and it was a wonderful time. My strongest memory from the event is of one of the morning Bible teaching sessions: we were deep in the word of God, when suddenly there was a commotion at the back of the room – someone had heard a particular birdsong out the window! The poor Bible teacher had to stop while folks clambered over chairs to lean out, and much excited chatter ensued around whether it was a fairly common Blackcap or a Melodious Warbler, which was possible in Portugal but an extremely rare sight for many people at the forum. Only at an A Rocha meeting could a whole plenary session be brought to a standstill because of a particular bird possibly being outside the window!

I was brought up to love birds and I can identify probably the top fifteen to twenty English ones, but I'm no expert. At A Rocha, though, I was introduced to the strange and wonderful world of bird watching (and to the strange and wonderful people who inhabit that world), and I learnt the excitement of seeing a bird I had never glimpsed before, and of standing still and quiet in a field or wood and learning to distinguish the different calls I was hearing.

In this chapter, at long last, we get to living creatures. We will look in our next and final chapter at land creatures – including the human land creature, of course, and at what it means to be made in the image of God – but here in Chapter Five, we focus on the sea creatures and the birds of Genesis 1.20–23, seeing how they feature in the Bible and what we can learn from them today.

Waters teeming and birds flying

Finally the spaces are ready to be inhabited! There is light, atmosphere and water; land and vegetation; warmth, and a rhythm of day and night and seasons. Now the world can sustain animals, and the first to be created are the creatures to fill the waters and seas, then the birds to fly in the air.

As we saw in the last chapter, so here too the picture is one of fabulous abundance, with seas and skies teeming with life. We would get a glimpse of what this would look like if we have had the privilege of swimming in pristine coral reefs with their incredible multitude of colourful fish, or witnessed footage of mass herring shoals, sometimes numbering tens of millions, covering several square miles. Or we may have been lucky enough to have seen a vast starling murmuration. The largest flock of birds ever caught on camera was in the BBC's *Planet Earth* series where they filmed red-billed quelea swarming over the African Savannah.[2] On a smaller scale, I remember holidaying as a family one year in a Mongolian tent called a yurt in woodland in the Cotswolds, England. The surrounding seventy acres had been set aside as conservation land, and as we walked through the wildflower meadows, clouds of butterflies flew up around us in numbers I had never experienced before.

We revel in the diversity of what God has created in the seas: the tiny pale-brown Paedocypris progenetica fish, found only in the Southeast Asian islands of Borneo, Bintan and Sumatra; the brightly coloured flouncing Siamese fighting fish; the inflatable puffer fish; the bizarre blob fish and the giant oarfish, which can grow to 11 metres long. And we revel too in the diversity of what God has created to fly in the skies: the tiny bee hummingbird; the oddly crested Andean cock of the rock; the stunning keel-billed toucan and peacock; the mighty wandering albatross and the ostrich. To say we are merely touching the tip of the iceberg here is an understatement: there are something like 11,000 bird species and maybe as many as 30,000 fish species, though no one knows for sure. And when it comes to insects we are even less certain, with some estimates at between six to ten million. As British biologist J. B. S. Haldane is reputed to have said: God, if he exists, must have 'an inordinate fondness for beetles'![3]

As we have seen in previous days, God looks and sees that it is good. One gets a sense that the creation, with its colour, vibrancy and diversity brings incredible pleasure to God. This is no dispassionate God who thinks, 'This is okay; it will do for the time being', but a God who looks at what he has made and views the teeming shoals and swarming flocks with deep satisfaction.

God not only sees that it is good: for the first time in the text God pronounces a blessing on his creatures telling them to be fruitful. The implication is that he has made the beginnings of the myriad of sea and sky creatures that we see today, and with his blessing God sets them off to multiply and fill the spaces he has created for them.

It is good for us to note this because although we are familiar with God's blessing on Adam and Eve to be fruitful and fill the earth, we sometimes miss the fact that that God gives that blessing to all his creatures. It is also worth considering that the phrase, 'living creatures', in verse 20 is the same as that used of 'the adam' in Genesis 2.7, where it says that God breathed the breath of life into his nostrils and 'the man became a living being'. Sometimes people ask if other creatures have souls, and the Hebrew word for 'creatures' and 'being' is the same word: *nepeš*, which elsewhere is translated 'soul' but simply means 'being' or 'life' (eg. Deut. 6.5, 'Love the Lord your God with all your heart and with all your *nepeš* and with all your strength', and Ps. 103.1, 'Praise the Lord my *nepeš*, and many other places). So both Genesis 1.20 and 2.7 use the same word yet one is translated 'creatures' and the other 'being'. In fact the KJV is more blatant, translating 1.20 as 'the moving creature that hath life' and 2.7 as 'living soul'. (It is interesting to note too that the literal translation of Prov. 12.10 is 'a righteous [person] has regard for the *nepeš* of his animal', which the NIV renders 'the righteous care for the needs of their animals' – what a bland translation!) The literal translation for both sea animals and human beings is 'living soul' or 'living being' and we must be careful not to ascribe something called a soul to humans but not to other creatures: biblically everything that has the breath of God in it has – or indeed *is* – a soul.

In the verses that we are considering in this chapter we pick up again the contrast with the Babylonian creation story, *Enuma Elish*, and the Genesis author's desire to de-divinize the natural world. In Chapter One, we saw that the Babylonian narrative depicted the creation of the world as coming from the body of the monster sea goddess, and the seas as representing evil and chaos. But here in the Genesis text, along with the sun, moon and stars of Day Four, the great creatures of the sea are created together with everything else. The actual phrase 'God created' is only used three times in the text: right at the start in 1.1; when God creates people in 1.27, and then here in 1.21. Maybe in this way the author emphasizes that, rather than representing the ultimate face of evil, the sea creatures too have been created by God. And, not only are they created, but they also are blessed and are seen as good.

Look at the birds of the air

Although on Day Five the water and sea creatures are created first, we will look at them last and think first about birds. In fact, watching the birds is a very biblical thing to do because Jesus, of course, instructed us to 'look at the birds' (Matt. 6.26). It is an instruction that my A Rocha friends take great delight in – Jesus is telling us all to become birdwatchers (and botanists to study the lilies of the field)![4]

I am sure we are getting the picture now that, while we tend to focus on people when we read the Bible (understandably so), if we are prepared to widen our gaze then we see that the wider natural world is also clearly part of the story. Some 30 or so different types of bird are referenced in the Bible (sometimes we're not sure to what exact species the Hebrew refers), and in Jewish tradition they are mentioned even before the very act of creation. The Jewish texts called the Talmud suggest that the Spirit hovered over the waters like a dove.

The dove is one of the best-known birds of the Bible and is almost the first fully referred to there, beaten by just one verse by the raven. Those birds were the two that Noah sent out from the ark to see if the water had receded enough and if there was dry ground. The dove becomes the hero (heroine?) of the story, returning with an olive branch in its beak and therefore the good news that vegetation was appearing. It is interesting to note that the Babylonian story of the flood (called The Epic of Gilgamesh) also has a dove being sent out of the boat by the main character, Utnapishtim. While many of us probably see a white dove in our minds, that is actually a later European variety: biblical doves were turtle doves or rock doves, more similar to today's urban pigeons.

Doves are a central element of the Old Testament sacrificial system and are used in general as burnt offerings, but also as purification after childbirth (Lev. 12, which we see Mary practising in Luke 2.24), and Jesus drives out the temple stallholders who were selling doves, amongst other things (John 2.13–16). Doves were easy to keep in captivity and so could be offered by those who could not afford to sacrifice a larger animal, and they were part of the list of birds regarded as 'clean' and therefore edible in Leviticus 11.14–19.

Alongside the ark narrative, the other significant place a dove appears in the Bible is at the baptism of Jesus, where the Holy Spirit descends on Jesus 'like a dove'. Linked with the opening verse of the Bible, the Holy Spirit has long been symbolized as a dove, especially in art. You may know the piece by the 1960s American pop art artist Andy Warhol, called, 'The Last Supper (Dove)', which juxtaposes Da Vinci's classic painting of the Last Supper with the modern personal-care product brand Dove, whose logo hovers over Jesus' head.

In rabbinic tradition doves represent faithfulness. A symbol of the beloved in the Song of Songs, they were believed always to be faithful to their partner, as God is faithful to Israel. And, as the dove returned to Noah, so Israel is called to return to God. As the dove 'found no resting for the sole of her foot', so the Jewish people have often found no rest among the nations.[5]

After its role in the flood story, the raven is also referred to elsewhere in the Bible, and is the bird chosen by God to feed Elijah (I Kings 17.2–6). Elijah has just prophesied to king Ahab that, as judgment, there would be no more rain or dew for the next few years, and he has to flee to escape Ahab's anger. He hides in a ravine at the edge of the land of Israel, and there God provides for him: water from a brook and bread and meat brought to him by the ravens. This may sound romantic, but I dread to think what the meat brought by ravens would have been like, and I hope Elijah was able to build a fire to cook it! As with Hagar and water, so through birds we see again God as provider, Jehovah-Jireh – something demonstrated also in Exodus 16 when he provides for his people by sending a big flock of quail in the early evening.

A variety of different birds are used metaphorically in the Old Testament, but one of the best known is the eagle:

> Do you not know? Have you not heard?
> The Lord is the everlasting God, the Creator of the ends of the earth.
> He will not grow tired or weary, and his understanding no one
> can fathom.
> He gives strength to the weary and increases the power of the weak.
> Even youths grow tired and weary, and young men stumble and fall;
> but those whose hope is in the Lord will renew their strength.
> They will soar on wings like eagles; they will run and not grow weary,
> they will walk and not be faint. (Isa. 40.28–31)

I remember a time when my children were young and the family finances were really tight. We didn't know what was going to happen in the immediate future and I felt very insecure, as if I was being dangled over the rocky face of a canyon and at any point might fall and crash to the bottom. In a time of prayer, I gave my feelings to God, and he expanded the picture to show me that, though I felt like I was being dangled over the canyon, actually my hands were being held securely in the grip of an eagle who was flying me over the rocks, navigating my path and keeping me safe. It was uncomfortable being held like that and flying with the eagle, and I would much rather have been put safely down on the ground . . . but I knew I was not going to be dropped and that in its grip I was secure.

The Gospel of Thomas (and also the Qur'an) tells a story about Jesus when he was five years old, taking water from a brook, mixing it with the soil and making twelve clay sparrows. He gets told off by his father because he is doing this on the Sabbath and, in response, he claps his hands and the sparrows come to life and fly off. I would love this story to be true but sadly it is almost certainly invented. However, in the gospels, Jesus does affirm his heavenly Father's love for sparrows when he talks to his disciples about not being afraid. Sparrows are everywhere and are cheap – they are sold two for a penny or five for two pennies (Luke seems to have been better at bargaining in the marketplace!) – and yet even with this bird that is so prevalent as to be hardly worth noticing, 'not one of them will fall to the ground without your Father's care', and, 'not one of them is forgotten by God' (Matt. 10.29–31; Luke 12.6–7). Arguing from the lesser to the greater, if this is the case with sparrows whom God loves, then just think how much your heavenly Father loves and cares for you!

You may well know this little ditty by Elizabeth Cheney which expresses the sentiment wonderfully:

Said the robin to the sparrow,
'I should really like to know,
Why these anxious human beings
Rush about and worry so.'
Said the sparrow to the robin,
'Friend I think that it must be,
That they have no Heavenly Father,
Such as cares for you and me.'

One other way Jesus shows his awareness of the birds around him is through his anguished outburst about Jerusalem: 'How often I have longed to gather your children together, as a hen gathers her chicks under her wings' (Matt. 23.37). He sees the trouble that is brewing for Jerusalem and yet she will not listen and repent. I expect we can picture this in our minds, but sadly today the vast majority of chicks will never know what it feels like to be under their mother's wings, and many of us reading this will never have seen a hen protecting her young in this way: our only experience of hens or eggs is either chicken wrapped in plastic or eggs sitting in a box on the supermarket shelf.

Chickens feature too in the final events leading up to Jesus' death as the cockerel crows three times. Poor Peter realizes that Jesus' words to him have come about: he has run away and denied Jesus at the very time when he should have been standing with him. Linking together birds and fish, the two subjects of this chapter, it is later on, after eating fish together, that Peter experiences Jesus' resurrection grace and forgiveness (John 21.10–19).

There is much in the Bible about learning from the natural world, and there are many things we can absorb from birds. In John Stott's beautiful book, *The Birds Our Teachers: Biblical lessons from a lifelong bird-watcher*, he says humorously that he has developed a new branch of science called 'orni-theology'.[6] He takes eleven birds of the Bible and looks at what we can learn from each one in lessons about faith, repentance, self-esteem, freedom, work and other things besides. Through it all he reflects Martin Luther's view from his commentary on the Sermon on the Mount in 1521 that God 'is making the birds our school-masters and teachers . . . In other words, we have as many teachers and preachers as there are little birds in the air'.

I have just come indoors from sitting outside to eat my lunch in the back garden. Though we live in an urban area with only a moderately sized garden, we have worked hard to make it a place that is friendly for other creatures to share with us. As a result we have a flock of sparrows and a good number of other birds in and around about. As I've been eating, I've been watching the sparrows as they've chatted together in the bushes, splashed about in the pond and had a dust bath on a dry patch of ground. There is a robin who I think must be feeding a second batch of youngsters, as each time I see him he has food in his mouth; and he has been hopping around near me, watching me with his bright eyes, head on one side. I've been reflecting on the little conversation between the sparrow and the robin that we read and on how much my heavenly Father cares for me and will provide for me. It is a good thing indeed to let the birds be our teachers.

As we do so, we can join them in praising God. I remember one evening when I was leading the main worship meeting at Spring Harvest (a large Christian event in the UK). There were five thousand of us in the Big Top tent, singing to God, and when I stepped outside briefly, I heard a blackbird in a tree, also singing its heart out. I don't quite want to say that he was singing to God in the same way that we were, but his beautiful song reminded me that the whole creation was also praising God in its various ways (Ps. 148.10). As we saw in our last chapter, the vision of heaven in Revelation 4 includes the four living creatures around the throne, one of whom 'is like a flying eagle'.

Avian emergency

Birds are amazing creatures. There is the colourful budgerigar who is the only bird species so far that has been found to 'catch' yawns – maybe a way of showing empathy or group awareness. There are the chicks of the hoatzin in South America that have a claw on each wing for the first three months of their lives. Their nests are built above water so, if the chicks fall out, they can use their claws to grab onto a branch and haul themselves out of danger and back up the tree (the hoatzin as a species is also known as the 'stink bird' because of their terrible smell!) The Bassian thrush from Australia hunts by directing its farts at piles of leaves which makes the worms move around so the thrush can see where they are. And the woodpecker stores acorns in individual holes bored into the trunks of trees – they have been known to hide up to 50,000 acorns in a single tree, each one in its own hole.[7] And who of us, once they have seen it, could forget the amazing bowerbird with his elaborate courting display and carefully constructed nest decorated with all sorts of trinkets?

But birds are facing a crisis, with 14 per cent at risk of extinction and overall numbers plummeting worldwide. In the UK there are now half as many in the countryside as there were forty years ago. Returning to the humble sparrow, the UK tree sparrow has declined by a staggering 95 per cent since 1970 (and the house sparrow by 70 per cent).[8] Around the world, some of our most familiar birds are in trouble. The yellow-breasted bunting has declined by 90 per cent since 1980; the African grey parrot is now endangered; vulture populations in South Asia declined by around 95 per cent between 1993 and 2000 and are now beginning to disappear across vast areas of Africa, and the European turtle-dove is now classified as 'vulnerable to extinction'.[9]

I'm sure by now it will not surprise you to learn that habitat loss and climate change are two of the key drivers behind the decline in bird numbers. Of these, habitat loss is the bigger issue due to urban development and changes in agricultural practice. The desire for increased efficiency has led to hedgerows being destroyed to create larger fields with ploughing closer to the edges of fields, and damp areas and wetlands being drained to provide more land for production. Changes in crop practices have resulted in the land being in constant use, with autumn sown cereals meaning the land has no time to rest and recover over winter. The move away from mixed farms to ones that specialize in either arable or livestock production has meant there is no longer the diverse habitat that birds need to survive, and the increased use of pesticides and fertilizers has destroyed the insect population the birds feed on and made the ground sterile and poisonous. (The decimation of the vulture in South East Asia, for example, has been largely the result of poisoning from livestock carcasses contaminated with the veterinary drug diclofenac – a painkiller for sick livestock, but lethal to vultures, while licensing of that same drug in some European countries could destroy hard-won conservation efforts.[10]) When you add to this the way farm buildings have changed so many no longer provide space for nesting birds and bats, it is easy to see why our bird populations have declined so much – all around the world.[11]

Changes in farming also mean that the majority of birds in the world are now, in fact, factory chickens: there are three times as many domesticated poultry as wild birds.[12] Standards vary between countries and regions, but the vast majority of the chicken we eat comes from birds that have been kept in terrible conditions.[13] Bearing in mind God's love for his avian creation, I have come to the stark conclusion that it is simply unchristian to eat meat from chickens kept in this way.

Returning to wild birds, research in the UK has found that plant, animal and insect life is more abundant on organic than non-organic farms. The former are home to 30 per cent more species, and some endangered farmland species were found only on organic farms. There were 44 per cent more birds in fields outside the breeding season and endangered birds such as the song thrush were significantly more numerous on organic farms. In particular, there were more than twice as many breeding skylarks.[14]

One of the best things we can do on a personal level then is support farmers who are growing their crops and rearing their animals without using a lot of pesticides and fertilizers. The simplest way to do that is to buy organic produce where we can, but there are other non-organic schemes that are positive too, such as the LEAF (Linking Environment and Farmers) marque in the UK. These schemes are about 'integrated farm management', which includes the use of traditional techniques along with modern pesticides and aims to minimize environmental impact.[15] Organic produce is often more expensive – and for good reason – so I recommend switching one of your main food items to organic and then when that feels normal, switching something else, and so on.

Saying Yes to Life is being read by people from many different countries, and farms and farming methods vary hugely around the world. But, in the UK at least, farming is changing, and the UK farming community recognizes the need to stop harmful practices. Traditionally farmers have always sought to hand the land on to the next generation in a better state than they inherited it, but they need the support of consumers and policy makers to do this.[16]

Another driver behind the crisis is the hunting and trapping of birds, both for pleasure and as a means of protecting the smaller birds they prey on, which are also shot for sport. There is a sobering passage in a book called *The Highland Notebook; Or, Sketches and Anecdotes*, written in 1843, in which the author, Robert Carruthers, describes the measures taken by the manager of the Glengarry estates in Scotland to keep down the 'vermin' which prey on the grouse. The landlord (an

Englishman called Lord Ward) was annoyed that he was losing his game, and so he 'set about a vigorous system of war and extermination against all his vermin intruders', giving prizes to the gamekeepers who were most successful. In three years they successfully destroyed 'four thousand head of vermin' and were pleased to see their stock of game increase as a result. Among the predators were 27 white tailed sea-eagles; 15 golden eagles; 18 osprey; 98 peregrine falcons; 275 kites; five marsh harriers; 63 goshawks; 285 common buzzards; 371 rough-legged buzzards; three honey buzzards; 462 kestrels; six jer falcon toe-feathered hawks; nine ash-coloured hawks; 1421 crows; 475 ravens; 35 horned owls; 71 common fern owls; three golden owls and eight magpies.[17]

Despite it being illegal, raptors are still persecuted today around the world. In New Zealand, a farm worker was found to have poisoned 406 wedge-tailed eagles through injecting chemicals into the necks of lambs which were left out to be eaten by the eagles, and in Argentina, 34 Andean Condors were found murdered in a similar manner.[18]

One bird of particular concern in the UK is the hen harrier which, having been hunted to extinction in the British Isles in the nineteenth century, had been reintroduced and was making a slow recovery. However, population levels are declining due largely to driven grouse-moor management and the hen harriers being killed in order to protect grouse stocks. Due to this, conservation organizations are calling for shooting estates to be licenced, and for there to be greater powers to investigate suspected wildlife crime, and stiffer sentences and repercussions for the estates and their employees convicted of it. Many conservationists, such as wildlife presenter Chris Packham and ornithologist Mark Avery, go further and call for an outright ban on grouse shooting.

Falling from the skies

Migration is a truly remarkable phenomenon with birds taking epic journeys across continents. Five European rollers that were tagged by A Rocha France crossed the Mediterranean in one night, then took different routes across the Sahara, gathered near Lake Chad to refuel and moved on to their wintering grounds in Angola and Northern Namibia, 7000 km away from the south of France.[19] That is nothing, however, compared to the Arctic tern which can do a round trip of around 25,000 miles, or the sooty shearwater which can do some 42,000 miles! We had the amazing experience of waking up one autumn Sunday morning to find the local common swift population had chosen the back of our house as a gathering point before setting off on their migration. Hundreds of birds were flying around and clinging to the back wall, cables and window frames (some came through the open windows into our bedrooms), until suddenly . . . woosh . . . they were gone, off on their journey to Africa!

But migration puts birds in terrible danger, and the greatest comes, tragically, from us humans. Horrendous slaughter takes place across Europe, Asia and Africa, as people come out to trap and shoot down the birds, whether for fun, food or pest control. Literally millions of birds are killed by guns, nets, traps or even glue put on trees.[20] The 90 per cent decline of the yellow breasted bunting, mentioned earlier, has been in large part due to it being hunted in this way in China.[21]

In all EU countries there is legislation to protect wild birds, but it is not enforced properly and penalties are not enough of a deterrent. Outside of the EU, many countries do not have any legislation at all. If you would like to explore this, the further resources section online has good organizations you can contact. The most useful thing is to find and support an organization in your country. Additionally you may want to consider avoiding travel to countries that are particular hotspots of illegal killing, such as Malta and Cyprus (and write to the tourism department to let them know your reasons).

Thinking back to the vision of heaven in Revelation 4.7 with the 'one like a flying eagle' worshipping at the throne of God, how sad that eagles are one of the bird species being hit particularly hard. Let us give the final word in this section to John Stott. He reflects on Jeremiah who saw the evils of habitat destruction: 'I looked at the earth, and it was formless and empty; and at the heavens, and their light was gone . . . I looked and every bird in the sky had flown away' (4.23–25). It is a warning of a possible return to pre-creation chaos and Stott says in response, 'Let's resolve to do all we can to protect and preserve our unique God-given environment, and so continue to enjoy its God-given biodiversity, not least its fascinating birds.'[22]

Every living thing with which the water teems

I have an ocean angel. He sits on my shoulder whenever I am writing or speaking about caring for God's world, and reminds me that the earth is 71 per cent water; that billions of people depend on the seas for their main source of protein, and therefore that when we are thinking about issues of poverty and environment, we must be careful to remember the seas as well as the land.

My ocean angel is called Dr Bob Sluka and he is a marine biologist who heads up A Rocha International's marine programme. Okay, he doesn't actually sit on my shoulder, but he has been very helpful in encouraging me always to include the seas, the ecosystems and species that live in and around them, and the people who depend on them.

In this chapter we are looking at Day Five and the creation of the animals to inhabit the spaces of sea and sky. God makes the water teem with 'living beings' and he creates everything that moves in the sea, blessing it all and telling it to increase and fill the water. Psalm 104 declares,

> There is the sea, vast and spacious,
> teeming with creatures beyond number –
> living things both large and small.
> There the ships go to and fro,
> and Leviathan, which you formed to frolic there. (Vv. 25–26)

In Chapter Three we met Jocabed Miselis from the Gunadule people off the coast of Panama. They view the world as having been created by two supreme beings, *Baba* (Big Father) and *Nana* (Big Mother), and the created world is said to be a reflection of their singing. Jocabed says, 'When you go to the sea and see the fish swimming and the dolphins jumping, they are dancing because they are rejoicing in *Baba* and *Nana* singing.'[23] As Christians, we affirm there is one creator God.

Yet can we learn something from the Gunadule people? Readers of this book are from different countries – some of us might live near the sea or ocean and gaze at it nearly every day; for others, seeing it will be rare. Whatever our experience, when we're near the ocean, can we regard the amazing creatures of the seas as rejoicing in their maker, reflecting his glory in their abundance and brilliance?

The seas and fish do not feature in the scriptural story in quite the same way as other elements of the natural world we have looked at in *Saying Yes to Life*. In the Old Testament, the people of God are land-based nomads, pastoralists and then urban and rural agriculturalists. It is other nations around them who are sea-farers, and the Israelites trade with them and enjoy the goods that come from across the water, but they themselves stay in the land that God has given them.

Nonetheless, the sea still plays its part. Most obviously, one of the pivotal events for God's people is the escape out of Egypt and the crossing of the Red Sea. As they leave Egypt and its tyranny, a new nation begins to be formed. Some time before this story happens, Moses encounters God in the burning bush. God tells Moses:

> I have indeed seen the misery of my people in Egypt. I have heard them crying out because of their slave drivers, and I am concerned about their suffering. So I have come down to rescue them from the hand of the Egyptians and to bring them up out of that land into a good and spacious land (Ex. 3.7–8).

These verses are amongst my favourite in the Bible. They are full of verbs: God sees, hears, is concerned, and comes down to rescue them. The whole story of the Exodus tells of an active God, who takes the initiative in bringing salvation to his people. And at no point is this more true than when the Israelites find themselves with the uncrossable sea in front of them and the chasing Egyptian army fast coming up behind. As Moses stretches out his hands, God sends a mighty wind that divides the sea and drives it back into two walls. The exposed sea bed is turned into dry land and the people cross, leaving the Egyptian army to drown as God sends the sea back into its place. No wonder Moses and the Israelites sing:

By the blast of your nostrils
  the waters piled up.
The surging waters stood up like a wall;
  the deep waters congealed in the heart of the sea . . .
Who among the gods
  is like you, Lord?
Who is like you –
  majestic in holiness,
awesome in glory,
working wonders?
(Ex. 15.8, 11)

The other well-known story that features the sea is at the other end of the Hebrew Scriptures: the story of Jonah. Poor Jonah is rather a hapless character. He does not want to follow God's word to go to Nineveh, so gets on a ship going in the opposite direction and finds himself thrown overboard and rescued by being swallowed by a big fish. When he finally obeys God and arrives in Nineveh, much to his annoyance the people repent and God does not bring the judgment Jonah had been looking forward to watching. We do not even know what happens to Jonah. We are simply left with him sitting in anger under a dead plant with the Lord telling him that, as God, he has every right to shown concern for the city with its more than 120,000 people – 'and also many animals' (Jonah 4.11). For all its oddity as a book of the Bible, this text gives a beautiful insight into God's heart for people (and animals). Though he is a God of judgment, he is also a God of love and mercy, and his desire is always to show compassion.

Beyond these two stories, the sea is closely linked to the temple.[24] The temple was regarded as the meeting place of heaven and earth, and represented the two coming together. It was designed to display many features of the natural world, but also contained God's presence in the form of the Ark of the Covenant. Solomon's temple included a sea 'of cast metal, circular in shape, measuring ten cubits from rim to rim and five cubits high' (1 Kings 7.23). It stood on twelve bulls, held about 44,000 litres and must have been very impressive! Ezekiel's vision reflects the importance of the sea, as water flows from the temple into the Dead Sea, bringing it to life so that it is filled with large numbers of fish – plenty for everyone to catch (Ez. 47.1–12).[25]

In the New Testament Paul travels a great deal by sea as he goes on his missionary journeys, taking the good news of Jesus throughout Asia Minor and into Europe. Much of Jesus' ministry is around the Sea of Galilee (strictly speaking not actually a sea, of course, but a large lake), and those he speaks to are embedded in a fishing culture and live off the lake's resources. We see him calling Simon Peter to follow him and fish for people, having given him an amazing catch of fish when he and his boat mates had been labouring all night and not caught anything. We also observe Jesus calming the storm and calling Peter out to walk on the water towards him. We read of him miraculously feeding thousands of people with just five loaves and two fish. And, as we noted earlier, it is by the shore, having cooked and eaten a breakfast of fish together, that the resurrected Jesus forgives Peter for his denial of him, and calls him now to take care of Jesus' followers, his 'sheep'.

Apart from the Leviathan, we do not get descriptions of different species of fish (as we did of birds) in the Bible (the only differentiation given is in the lists of clean and unclean food, where there is a division made between those that do and do not have fins and scales, seen in Lev. 11.9–12 and Deut. 14.9–10). Nonetheless, fish take their place alongside the whole creation. They too suffer the consequences of people's sins. It is because of the behaviour of the people of God that 'the land dries up, and all who live in it waste away; the beasts of the field, the birds in the sky and the fish in the sea are swept away' (Hos. 4.3). Yet, those who live in the sea join with the whole community of creation in praising God. The Psalmist commands them: 'Praise the Lord from the earth, you great sea creatures and all ocean depths' (Ps. 148.7, and see also 69.34, 96.11 and 98.7), and in the vision of heaven that we have looked at already, we see not only the four living creatures but 'every creature in heaven and on earth and under the earth and on the sea, and all that is in them', singing 'praise and honour and glory and power' to the one who sits on the throne and to the Lamb (Rev. 5.13).

Theomoana

In our earlier section on birds, John Stott's 'orni-theology', revealed how we can let the birds be our teachers. Former Archbishop of Polynesia, Winston Halapua, who we met in Chapter One, has based his thinking on the notion of *theomoana*. *Moana* is the ancient Polynesian word for ocean and a term that is still used in many parts of Oceania today. Oceanic people have different words 'for the sea surrounding the land, the sea over the reef, the sea over the sand, the sea that drops from the land into deep waters, the sea that flows into mangroves . . . [and] the coastal sea with big waves for surfing and the lagoon'. But, *moana* 'speaks of the mystery of the depths of the sea'.[26]

Having different names for different aspects of the sea reminds me of the many names for different types of snow used by the Sami people of Finland, Norway and Sweden, and of the different names given to the river Ganges that we read about in Chapter Two. I only have the word 'sea' and 'ocean' in my vocabulary and I am challenged to take more notice of the sea and pay attention to its different forms.

Winston Halapua reflects on the connectivity between *moana* and *fornua*, 'land'. There is a strong bond between the people and the land and seas that support them, neither of which is owned but both of which need to be looked after and respected. We may think of the oceans as dividing us into continents, but actually all the oceans are connected, so *moana* 'holds the good news that all creation is interconnected. Each component in the atmosphere, in the ocean, on the land, finds its origin, definition, purpose, completion and continuity in relationship. Life in relationship is the essence of the moana and all its rhythm'.[27]

Hillsong's 'Oceans (Where Feet May Fail)' is a worship song based on Peter stepping out of the boat, putting all his trust in Jesus. Listen to these words that will be very familiar to many of us and unknown to others:

You call me out upon the waters
The great unknown where feet may fail
And there I find You in the mystery
In oceans deep
My faith will stand

And I will call upon Your name
And keep my eyes above the waves
When oceans rise my soul will rest in Your embrace
For I am Yours and You are mine

Your grace abounds in deepest waters
Your sovereign hand
Will be my guide
Where feet may fail and fear surrounds me
You've never failed and You won't start now

Spirit lead me where my trust is without borders
Let me walk upon the waters
Wherever You would call me
Take me deeper than my feet could ever wander
And my faith will be made stronger
In the presence of my Saviour.

These words would resonate well with the ancestors of the Oceanic people who made epic voyages across the ocean, and with the early Celtic Christians such as St Brendan as they set out in their coracles (the traditional small Celtic boats) to take the good news of Jesus to other lands, and it resonates well with us too in the journeys of faith we take in our own lives. As Winston Halapua says, 'We too are called to . . . adventures in faith as we move towards the unseen future'.[28] Reflecting on this chapter this Lent, could you pause a moment and think back to the journeys of faith you have taken and to any that you feel you are currently on? Thank God that he has kept you safe thus far and ask him to continue taking you out into the deeper waters where you have no choice but to trust in your Saviour.

Revealing the hidden things of God

Most of us, when we look at the seas or the oceans, can hardly begin to imagine what lies beneath, but marine biologists know there is an incredible, diverse world increasingly being discovered. Christian marine biologists see their work as 'revealing the hidden things of God' – ultimately 'to glorify our Father in heaven'.[29]

The oceans are the least explored places in the world with wonders that a lot of us will never experience, so let's get a glimpse under the waters through this account from my ocean angel, Bob. He made a dive to research whether a national marine park in the Caribbean was making any difference:

> Our job on this day was to swim out through the coral reef and go down over the edge of a very deep trench called the 'Tongue of the Ocean'. We wanted to dive down to 100–110 feet and see what was there and if there were any bigger fish. So we swam out over beautiful coral canyons through crystal clear water and I went off on my own (which you should never do!), hanging out over water which is several thousand feet deep. I'm looking at this wall, with waterproof paper and a pencil and clipboard, and counting fish.
>
> Of course you've got all this deep blue behind you and you're wondering . . . You get that feeling every once in a while, and you can only turn around very slowly because you're like an astronaut as you're neutrally buoyant. Then one time I looked down just below me. I could see a couple of hundred feet down, and I just saw something. I was trying to figure out what it was and I kept an eye on it through the rest of the dive. Then as I looked down one time I noticed that it was getting bigger, and it just kept getting bigger, and then something shot past me! I looked around feeling disorientated and I came face-to-face with this huge fish, as big as me, with big teeth.
>
> It was a barracuda, and they have this unnerving habit of coming right into your face and looking at you. As it breathes it opens its mouth because it is sucking in the sea water and passing it over its gills. I'm looking deep into its big eyes, he's looking at me, and of course I'm a bit afraid. But then I also thought, 'Okay it's a barracuda, it's clear water, this thing is not going to eat me – it's not even going to bite me'.
>
> Then my fear turned into a feeling of awe and wonder. The feeling that captured me was the beauty of this animal, and the

> beauty of the situation. As I sat there I thought to myself, 'This is something worthwhile to do. Here I am, I get to be out here studying and protecting these creatures.' Beauty was a huge driver in my desire to protect and study these animals, and beauty has been something that has drawn me to the Lord through the things that he has made.[30]

Moana in crisis

Our seas and oceans and creatures that exist in them are incredibly precious; they are needed for life itself. The water cycle, as we saw in Chapter Two, sustains life on the land. Our seas are a critical part of the climate system that distributes warmth around the world through large-scale ocean currents, and also of the carbon cycle, as they absorb CO_2 through the phytoplankton that live in their waters. They sustain, as we have seen, incredibly diverse ecosystems and vast numbers of species that together keep the oceans healthy. As human beings, we gain a huge amount from them culturally, and we depend on the seas for a lot of their resources.[31] In fact over three billion people rely on marine and coastal biodiversity for their livelihoods, which is why the focus of one of the Sustainable Development Goals (no. 14) is: 'Conserve and sustainably use the oceans, seas and marine resources'.[32]

At present we are not doing this well. The 2019 Global Assessment Report on Biodiversity and Ecosystem Services from the IPBES (Intergovernmental Science-Policy Platform on Biodiversity and Ecosystem Services) stated that only seven per cent of marine fish stocks are being harvested below the level of what can be sustainably fished. Sixty per cent are at capacity and 33 per cent are being harvested at unsustainable levels.[33] What that means, put simply, is that we are over-fishing: we are taking too many fish out of the seas and in a way that is causing significant damage. Bottom trawling comprises huge nets being weighted (sometimes with metal beams or hydraulic dredges) and dragged across the sea floor. It scoops up everything and simply tosses this back in once the intended catch (predominantly prawns/shrimp and bottom dwelling fish such as cod) has been sorted.

There are examples of local fishermen using homemade bombs to destroy an area of the sea so the dead fish float to the surface, and of using nets with small holes to catch young fish that have not yet grown to maturity and had a chance to reproduce.[34]

Industrial overfishing exacerbates poverty for coastal communities that depend on the seas. The west African coast has experienced this particularly and there is a constant battle with illegal industrial fishing fleets that exploit the rich fishing grounds, leaving dwindling stocks for local fishermen.[35]

Sharks and rays and coral reefs are of particular concern. There are over 1,000 species of sharks and rays and, incredibly, new species are being discovered every year.[36] But they are being horrendously hunted – predominantly for their fins and meat but also their oils and cartilage – to the extent that one in three shark and ray species is at risk of extinction.[37] Sharks are also endangered because, as commercial fishing depletes fish stocks, it reduces the sharks' available food too; sharks also get caught in the vast nets that are put out. An astonishing 100 million are killed every year.[38]

The dwindling shark and ray population warns us of the problems of overfishing. The state of our coral reefs alerts us, once again, to the impact of climate change. Currently, a third of reef-building coral species is at risk of extinction, but scientists are predicting that coral reefs will disappear completely this century if we do not keep climate change within the limits we have already discussed in previous chapters. This is serious because coral reefs harbour the highest biodiversity of any ecosystem globally and directly support over 500 million people worldwide, mostly in poor countries. They also form – as do mangroves – a protective barrier for the land against storms and tidal surges.

Alongside overfishing and climate change, the other issue that cannot be ignored when it comes to our seas and oceans and all the living things in and around them is pollution – and particularly plastic pollution. Over the last few years we have become frighteningly aware that we are pumping plastic into the seas in terrible quantities and we will all have seen shocking images of vast floating islands of plastic; sea birds feeding their chicks small pieces of plastic; turtles wrapped in plastic fishing nets, and other marine life found dead with plastic bags and other bits inside them. In the Philippines last year, a dead whale was found to contain 40kg of plastic, including 16 rice sacks and many shopping bags.[39] Plastic never goes away, it breaks down into small particles called microplastics which are being found in the stomachs of many different animal species and on beaches the world round. The raw material used to make plastic is also a microplastic called a nurdle or plastic pellet and these are washing up on beaches globally, having absorbed toxins, and contributing to our plastic pollution problem.[40]

The greatest amount of marine plastic pollution comes from the land: from blow-off from coastal communities and industry, and also from rivers carrying inland waste down to the sea. It is shocking to realize that three billion people currently have no safe way of managing their plastic (two billion of them have no waste collection at all), and that plastic piles up in rivers and waterways, causing serious health consequences to the (generally poor) communities who must live among it, before it eventually washes down into the seas and oceans.[41]

Blowing the conch horn

Winston Halapua tells of how when something of importance in the Pacific Islands needs to be signaled, a conch horn is blown. In his Letter for Creation to Archbishop Justin Welby, he says:

> We need to blow a conch to alert the world of danger not only to ourselves but to the whole planet earth. We need to call for a working together to care for our common home. We need to raise prophetic voices today. We face great crises and need to face them together. We have the opportunity to forward a new movement of caring more deeply for God's creation, of celebrating its wonders and of discovering our common humanity.[42]

How can we respond to the crisis that is unfolding in our seas and oceans?

One encouraging thing is seeing churches engaging in practical action. The Methodist Church in Fiji has partnered with WWF in working with their communities to observe a period of 'Tabu' during the Kawakawa (Grouper) spawning period, when the usually solitary living grouper gather in huge numbers at specific times and places in order to find mates. In addition, every Methodist is expected to plant four trees a year, and a number of those are being dug in along the shoreline, helping counteract the deforestation that has taken place there which impacts the coral reef by increasing the amount of sediment and pollution entering the normally clear water the corals need to thrive.

In Coqueiral, one of the poor areas of Recife on the coast of Brazil, the local Baptist church has become a leader in responding to plastic pollution. The River Tejipio runs through the area but has become piled high with waste – in some places a metre thick. When it rains the river rises because of the rubbish blocking it. One particularly bad flood caused hundreds of homes to be destroyed and Coqueiral residents became ill because the river was so polluted with sewage. The church opened its doors, inviting people to sleep on the floor, and this was the start of its work with the community and local authorities to change practices and clean up the river. As Pastor José Marcos da Silva says , 'I used to think it was God's problem and he had to stop the rain. Now I realize we had to get together with the people and take responsibility to act'. The church runs a handicrafts scheme to turn the plastic into products and has undertaken educational work in the community. A march through the city with 14,000 signatures has led the local authority to start taking action too. The river needs constant attention and changing attitudes is slow work, but every piece of plastic taken out of it – or not put in at all – is one less that will harm the people living there and eventually find its way to the ocean.[43]

The organization Christian Surfers hold beach clean-ups at their UK gatherings and also internationally, and have changed their plastic membership cards to cardboard discs. At Jesus Surf Classic – one of the longest running and biggest surf competitions in Britain – they have stopped selling bottled water and make sure there is only re-usable crockery available. They are also partnering with Surfers Against Sewage to push the government to take stronger action on climate change as they see the detrimental impact this is having on the spaces they love so much.[44]

Tubestation – a Methodist 'Fresh Expressions' church in Cornwall – has got very involved with their nearby beach. A few years ago, big winter storms washed up huge amounts of rubbish so they did a beach clean every Sunday after church for two months. They said to their congregation, 'Our sung worship has finished but if you want to carry on loving the community and worshiping God in a different way, come and join us', and about thirty people stayed on each week. Joff Phipps, one of their leaders, told me, 'That was the start of a love interest with the oceans going deeper than just from a pleasure perspective'.

They then joined with Surfers Against Sewage and the Polzeath Marine Centre to do beach cleans together. Each year they hold a special week when they focus on looking after our oceans, giving talks about things like climate change and where to buy fish, and they include a day of prayer for creation. This had led to the group becoming more environmentally active, and they now have a wildflower area to attract bees and butterflies, and a vegetable patch with the produce to give away. Joff says, 'We started with sharing the Gospel to the surfing community, but we learnt you can't separate that from caring for the environment. Loving God and our neighbours brings us also to loving our beaches and the whole environment'.

If your church is near the coast, why not get involved in a beach clean up too? One opportunity to do this is with the international beach clean that takes place on the third weekend in September each year (it can be really positive to join with and support what others are already doing). If, having read this chapter you feel that's too long to wait, just get something organized and get out there – preferably in place of a Sunday service so your clean up is an integrated part of your worship life as a church. It is a really good practical thing to do together and can lead to helpful reflections on God's love for his blue world.[45]

If you are not by the coast, your church can still take action by moving away from items of single-use plastic (such as coffee cups and disposable crockery) and swapping to re-usable ones instead. The Church of South India, in its Green Protocols, is encouraging all its churches to become plastic-free by serving food on banana leaves or using steel plates and asking people to wash up after use.

In our churches and as individuals there is action we can take on plastic. The conch horn has been blown and there is no longer any excuse for us to continue life as usual: we must all change our practices. Tearfund has developed a strong focus on plastic because of the recognition that it is harming billions of people as well as the marine environment – Tearfund's report, 'No Time to Waste: Tackling the plastic pollution crisis before it's too late', showed that one person dies every thirty seconds in the developing world from diseases caused by plastic pollution and other rubbish.[46] We are working with church communities in Pakistan, the Democratic Republic of Congo, Brazil, Nigeria and Haiti to help them deal with the waste problems they are facing, and to create livelihoods. Alongside these projects, we are running The Rubbish Campaign, calling on four multinationals – Nestle, Unilever, Coca Cola and Pepsi – to change their practices and reduce plastic pollution in developing countries. All of this will save people's lives, as well as reduce the amount of plastic entering the oceans. Africa is leading the way on plastic bags – at the time of writing, eleven African countries have banned them or placed taxes on them. But we are nowhere near tackling the problem comprehensively, and you and your church can join Tearfund's campaign, calling on businesses and governments to change their practices (see the further resources for this chapter at <www.spckpublishing.co.uk/saying-yes-resources>).

Individually, too, there is much we must do, and I hope the need to take action on climate change is a message that is resonating clearly in this book as we travel through Lent together. Hopefully you have already taken some action over these last few weeks, but we need to keep asking ourselves: what more can we do?

We must break our plastic addiction and take every step we can to move away from single-use plastics. In our food shopping, our cleaning and toiletries, nappies and toys, men's razors and women's period products, there are many changes we can and must make. I have been trying to reduce my plastic usage for a long time now, but in recent years have stepped up my efforts. It takes thought and a willingness to do things differently when we have allowed ourselves to take much for granted in our societies. But it can also be fun and I have loved experimenting, particularly with my personal care products.[47] I have colleagues at Tearfund who are way ahead of me in reducing their plastic usage, and they challenge me never to think I am doing enough and always to look for new things I can try.

Finally, for those of us who eat fish and seafood, it is crucial we do so sustainably. The Monterey Bay Aquarium Seafood Watch programme gives advice on what fish is good or not to eat, as does the Marine Conservation Society – and both have apps you can download to make it easy to check when you're shopping or about to order in a restaurant.[48] In the UK, look out for fish and seafood that carry the Marine Stewardship Council's blue fish label which certifies sustainable fisheries. You are either on the end of a chain of blessing or a chain of cursing – what you buy and how you eat impact those far off for good or ill.

The hummingbird and the fire

Although we have looked separately at birds and sea creatures, the two are closely linked, with what happens on the land impacting the seas, and vice versa. For example, protecting forests for birds reduces the land's watershed by lessening erosion, which then helps the sea by reducing pollution. So wherever we are and whatever we do, we must remember that the land and the seas interact.

As we draw this chapter to a close, let us return to birds with a story about a hummingbird told by Wangari Maathai, which was recited to her by a Japanese professor:

> The story starts with an enormous fire, which breaks out and rages through the forest. All the animals flee to the forest edge to watch – all, that is, except a tiny hummingbird. 'I will do something about this fire,' says the tiny bird. So it flies to the nearest stream and dives beneath the surface. Rising into the air, it carries a bead of water in its beak that it releases over the flames. The fire is huge, but over and over the hummingbird flies to the stream, returns with a droplet in its beak, and lets it fall onto the flames. Each time, the bird believes that this one drop might make the difference.

> The other animals – some with large trunks and large mouths like the elephant, giraffe, lion and leopard – laugh at the diminutive creature. 'What do you think you're doing?' they jeer. 'You're only a hummingbird. You can see how big the forest fire is. Do you think you're going to do any good at all?'. Without wasting any time and tired of their discouraging words and inaction, the humming bird turns to the other animals as it prepares to fly back to the river, and says, 'Well I'm doing the best I can!'[49]

We have faced some big issues in this chapter and as we look at how to respond, we could feel like that hummingbird: small, insignificant, vulnerable and not wanting to be laughed at. But let us do everything we can to take action on the issues we have considered and draw courage from these words of Wangari Maathai, a woman who herself was prepared to stand up and bring about change:

> Hummingbirds though we may feel ourselves to be, we nevertheless have to take our small beaks and carry that bead of water (that droplet of change) to where it is needed, and do it over and over again, notwithstanding the . . . odds . . . or indifference from those more powerful than us. Alternatively, we may encourage others to step forward and join us. We will never know until we leave a fixed state and give ourselves the energy to move into action. In the end, all we are called to do is the best we can.[50]

For discussion

1 What experiences have you had of seeing 'teeming' in the natural world? Reflect on those and take time – on your own or in a group – to appreciate those experiences and thank God for them.
2 In this chapter, we ponder the suggestion that it is unchristian to eat meat from chickens that have been reared in cramped and intensive conditions. It is likely this has implications for the majority of us reading this book. How do you feel about this? Do

you agree or not, and why?

3 Watch this chapter's interview, featuring Professor Meric Srokosz from the National Oceanography Centre, and use his expertise on the oceans to stimulate your own thoughts and discussions. You can watch the video at <www.spckpublishing.co.uk/saying-yes-resources>.
4 Do you have a favourite sea/ocean place to be, or favourite memory? If you do, sit still and allow yourself to go back to that place or memory and, as you do so, ask God what he might want to say to you.
5 Make a mental note of the different ways in which you use plastic. Maybe look around you or around your house and notice how much there is. What plastic things will you decide now to abandon or find an alternative for?
6 How could your church get involved with the topics this chapter considers?

Prayer from young person in Vanuatu[51]

O Jesus,
be the canoe that holds me in the sea of life,
be the steer that keeps me straight,
be the outrigger that supports me in time of great temptation.
Let your Spirit be my sail that carries me through each day,
as I journey steadfastly on the long voyage of life.
Amen.

6
Let the land produce living creatures and let us make humankind in our image
(Genesis 1.24–31)

[24]And God said, 'Let the land produce living creatures according to their kinds: the livestock, the creatures that move along the ground, and the wild animals, each according to its kind.' And it was so. [25]God made the wild animals according to their kinds, the livestock according to their kinds, and all the creatures that move along the ground according to their kinds. And God saw that it was good.

[26]Then God said, 'Let us make mankind in our image, in our likeness, so that they may rule over the fish in the sea and the birds in the sky, over the livestock and all the wild animals, and over all the creatures that move along the ground.'

[27]So God created mankind in his own image,
in the image of God he created them;
male and female he created them.

[28]God blessed them and said to them, 'Be fruitful and increase in number; fill the earth and subdue it. Rule over the fish in the sea and the birds in the sky and over every living creature that moves on the ground.'

[29]Then God said, 'I give you every seed-bearing plant on the face of the whole earth and every tree that has fruit with seed in it. They will be yours for food. [30]And to all the beasts of the earth and all the birds in the sky and all the creatures that move along

> the ground—everything that has the breath of life in it—I give every green plant for food.' And it was so.
>
> [31]God saw all that he had made, and it was very good. And there was evening, and there was morning—the sixth day.

While writing this book, we have been sharing our garden with a family of hedgehogs. We have had hedgehogs off and on over the years, but these last few months have given us our closest encounters yet. A mother had three babies in a nest she built under a piece of garden furniture, and we started to see her snuffling around in the garden in broad daylight. Then at dusk one evening, much to our delight, the babies came out too. They didn't see us as we stood stock still, and they trotted around our feet, foraging in the vegetation.

We put water and hedgehog food out every day and soon a pattern emerged: the mum would come out early evening while it was still light (an early breakfast in peace and quiet without the kids), then she would come back at dusk with the babies for them to feed as well. It became part of our regular routine too, each evening, to go out and watch them pottering about the garden.

The babies grew and, just recently, they went on their way to find their own territories. But it has felt a special privilege to be part of their lives and, hopefully, we have done something to help hedgehogs in general because whilst stable on a global level, in the UK they have suffered a massive decline. It is thought that around 30 million hedgehogs populated Britain in the 1950s, but now there are possibly only around a million left, meaning nationally they are at risk of extinction.[1]

Encounters like these are precious as they underline the connection between us humans and the wider animal world. In her 2016 Archbishop of Canterbury's Lent Book, *I Am With You*, Episcopal priest Kathryn Greene-McCreight says, 'There is no true humanity without other creatures of God'.[2] As we will also see in this chapter, there is no true humanity without other humans: a human on their own is not fully who they are meant to be. But humanity does not only exist within the human community: without the wider community of creatures that God has created we cannot be what we have been created to be and we are the poorer for it.

In this final chapter our focus turns to land creatures, both human and other-than-human. First we will consider land animals in general and then we will look at human beings; at what it means to be made 'in the image of God', and what that tells us about our relationship with one another and the wider creation. As part of this we will turn our attention to a topic we have touched on in almost every chapter – food. We will give further thought to what and how we should eat: a topic that seems very fitting for Lent, focused as it is on fasting!

Creatures of the land

So now the final created space gets filled – the land. In similar fashion to verse 11 where the land is told to bring forth/produce vegetation, now it is to bring forth living creatures. As we saw in the previous chapter, the literal translation is the same as that for the sea creatures: 'living beings'. So the land is to produce 'living *nepeš*'. This is demonstrated towards the end of the chapter where green plants are given to 'everything that has the breath of life in it' (v. 30). Literally this could also be translated as everything that has the 'soul of life' in it, and it is a beautiful reminder that the soul is not some distinct spiritual entity that only human beings have, trapped inside our physical bodies, waiting to be freed. It is a description of who we are as created beings, both human and non-human. All of us have God's breath in us.

In the categories of living creatures given in verse 24 we see the basic division into domestic animals, land-based 'crawling things' and wild animals. As with the creation of vegetation, sea creatures and birds, the land creatures are made 'according to their kinds'. This is an important phrase to the author, who uses it five times in just the two verses 24 and 25. It reflects the overall stress in the Hebrew Bible on appreciating and respecting the distinct nature of different kinds. The word is used again when Noah, in Genesis 6.20, is commanded to take two of every kind of animal into the ark. Rabbi Norman Solomon argues that God and the text are concerned with biodiversity and the preservation of each separate and distinct species.[3] In Chapter Three we looked at the fertility of the land and in Chapter Five at the many kinds of birds and sea creatures. Here too, this phrase calls attention to the number and variety of different species that are envisaged.

In other words, we share this world with the most incredible and wonderful mix of strange, colourful, funny, scary, cuddly, scaly, odd, tiny, huge creatures that we could ever possibly imagine! Who could have thought up the star-shaped mole of North America with its 22 little tentacles on the end of its nose that it uses to find food, or the tiny elusive primate called the tarsia of southeast Asia, with its huge eyes, ability to turn its head 180 degrees and super-long back legs which enable it to leap up to five meters from branch to branch? Who would make up the saiga antelope of Eurasia with its strange nose that comes down over its mouth, or the aardvark of sub-Saharan Africa with its long tongue and kangaroo-like ears, or the lion-tailed macaque of India, with its stunning silver-white mane and tail that ends in a tuft like a lion's? Who would think to put such tufty ears on the European lynx and who ever could have imagined the duck-billed platypus with its duck-like bill and beaver-like tail?! Wherever in the world we live, in the city or in the countryside, we have amazing creatures around us – even if they are not all as fancy as those just mentioned. Why not pause for a moment to think about the animals that live around you and give thanks to God for such an abundance of life?

Having created the land animals, God pauses to look once more at what he has made, and he declares it 'good'. This deepens the affirmation we have seen throughout Genesis 1 of the inherent value and worth of God's creation. And it is worth stating again: God loves each aspect of the created world for its own sake – each thing has goodness in and of itself in the eyes of God. I was once privileged enough to stay in a lodge in the Serengeti in Tanzania. The view from my balcony stretched for miles and, as the sun set, I stood out and watched a troupe of about 50 elephants wandering across the savannah. I was struck by my irrelevance to them: they had no idea I was there and their lives would continue without me. Of course, I am aware of the painful politics around safari parks and the deep interrelation between people and the non-human animals that live there. The reality is my life probably *did* have relevance to them in a variety of ways (not least through my park fees). Nonetheless, it was a humbling experience watching these majestic creatures with the breath of God in them, knowing that God saw them too and declared them to be good.

A giant fish and a talking donkey

As we have worked our way through Genesis 1 during the course of this book, we have seen how the Bible is not only a story about human beings but a story about the whole world – indeed the whole universe! To be sure, the story centres around people and God's unfolding relationship with them, but the wider natural world is never far away, and the biblical text is full of trees, birds, fish, fields, gardens, stars, insects, the sun and moon, flowers, seas, rivers, rain, clouds, wind . . . and the animals created on Day Six. In fact, animals in general are pictured alongside people right the way through the Bible.

Noah's Ark is an obvious favourite in this regard, and we touched on it earlier, but it is important to notice that God's covenant with Noah is actually also with 'every living creature that was with you' (Gen. 9.10). It is emphasized seven times that God's covenant is with 'all living creatures' or 'all life on the earth' (9.10, 12, 13, 15–17). Really, this ought to be called the Earth Covenant.

As David Clough points out, 'The people of Israel are always to be found in the company of . . . animals', particularly domestic animals but sometimes wild ones too.[4] The Patriarchs (Abraham, Isaac and Jacob) and their families always have animals with them, and it is interesting to note that on the night of the Passover, when the Egyptians are wailing at the death of their first born sons, even the Israelites' dogs keep quiet and do not bark (Ex. 11.7). As we mentioned in Chapter Three with reference to the land, animals are also included in the laws of the Sabbath, including the wild animals that may eat off the land in the Sabbath years (Lev. 2.6–7).

When we looked at Jonah in the last chapter, we saw that God spared the repentant Nineveh with its 120,000 people – 'and also many animals', and the story tells us that the animals as well as the people fasted and were covered with sackcloth (3.7–8). Animals are therefore seen as having their own capacity to act and respond to God.

In the Bible we see God speaking to creatures: the snake in Genesis 3; the raven who fed Elijah; the wild animals he told not to be afraid (Joel 2.22), and the fish that vomited Jonah out. In the amusing story of Balaam's donkey we find a creature far more aware of God than his master (Num. 22.21–35)!

Alongside the land, the animal kingdom will suffer the effects of judgment for people's sins (eg. Ez. 14.21; Zeph. 1.2–3; Joel 1.18–20). But, more positively, taking their place with all the other aspects of the natural world that we have looked at in *Saying Yes to Life*, land animals are called to join creation's choir in praising God. It is interesting to note that Psalm 65.2 in the NIV says 'all people will come' to praise God, but the literal translation is 'all flesh', in other words, all living things. The beautiful image of the whole creation praising God, that we have seen so many times already, finds its fulfillment in Revelation 4 and 5 with the four living creatures – one with a face like an ox and another like a lion, representing the domestic and wild animals respectively, alongside one with a face like an eagle and one with a face like a person – as the central worshippers, subsequently joined by 'every creature' in creation (5.13).[5] When we praise God, we do so in good and full company.

Like the Patriarchs, Jesus too lived his life in the company of animals. Although the gospel stories don't specify this, the fact that Jesus was laid in an animal feeding trough when he was born and that shepherds came to worship him, suggests there may well have been sheep and goats, cows and donkeys around the place where he lay.

There is an important little verse just before the start of Jesus' ministry when he is being tempted in the wilderness. Mark tells us that 'he was with the wild animals' (1.13). Following a long tradition of scholarship, which sees this passage as portraying Jesus as the second Adam who is tempted but this time does not succumb, Richard Bauckham picks up on the eschatological hopes of Isaiah 11.6–9, with its beautiful picture of concord within the animal kingdom (what scholars call 'the peaceable kingdom').[6] In this Isaiah passage, as Bauckham highlights, there is peace between wild and domestic animals, and between wild animals and people, made the more poignant by highlighting peace with the most vulnerable of animals: those that are young (yearlings and infants). In the wilderness, Jesus encounters three kinds of non-human beings: Satan, the wild animals, and the angels. Satan is only to be resisted. The angels are there to minister to him. In between are the wild animals: 'They are enemies of whom Jesus makes friends'.[7] Jesus being with the wild animals in this peaceable way not only confirms their independent value (he does not try to control or dominate them) but affirms that 'the kingdom of God inaugurated by Jesus includes the wider animal kingdom.[8]

This is a very important point for us to grasp and resonates with our discussions around eschatology in Chapter Four where we saw that the whole created order (rather than humans alone) is caught up in God's plans for redemption. As Indian theologian, Ken Gnanakan says, 'The promise of God cannot be restricted to people'.[9] We may therefore look forward to the transformed heaven and earth including other creatures and, yes, maybe even particular creatures we have known and loved in our lives, as well as the many we have not.[10]

Animals were naturally part of Jesus' world. He talks about sheep and wolves, snakes, dogs and oxen, and it is an animal – a donkey – that brings Jesus into Jerusalem towards his death. This is not to say that Jesus saw no differentiation between people and other animals: his saying about sparrows and his teaching on the Sabbath (see below) argues from the lesser to the greater, and assumes that sparrows, sheep and oxen are not as valuable as people, though, of course, that does not mean they have no value at all. We see this too in his allowing the demons to take possession of the pigs when teaching in the region of the Gerasenes (Matt. 8.28–32; Mark 5.1–13; Luke 8.26–33). In the understanding of the day, the demons had to go somewhere otherwise they would have returned to the man. So casting them into the pigs is a lesser of two evils in the time before Jesus had won his full victory on the cross, and a recognition that the pigs were of less value than the man. But, in the light of the broader biblical picture, we cannot say from this story that Jesus had no concern for animals at all.[11]

The Bible shows us a God who is deeply concerned for his creation, and not least for his animal creation. It is God who provides water for the beasts of the field and grass for the cattle; God who provides all creatures with their food so that they look to him and are satisfied (Ps. 104.11, 14, 27–28). It is God, in Jesus, who says that the Sabbath laws are there to serve his creation, not the other way round: if an ox has fallen into a pit or a well on the Sabbath day, of course the owner would pull it out and not leave it there till the Sabbath is over – that is part of what it means to 'do good' (Matt. 12.11; Luke 14.5). Therefore, as we will explore further below, it is expected that we too follow God in doing good towards what he has made and looking after the other creatures of this world.

The place of animals in the Bible shows us they are an integral part of the story of redemption; therefore we should remember their place in our lives as well. We see something of this happening in the *Thanksgiving Address* from the Native American Haudenosaunee nation, which goes through and addresses each part of creation – the animals:

> We gather our minds together to send greetings and thanks to all the Animal life in the world. They have many things to teach us as people. We see them near our homes and in the deep forests. We are glad they are still here and we hope that it will always be so.[12]

How might our church services – or our individual prayers – be changed if we paused and took time to remember the animals in our world; recognized their presence and were thankful for it; acknowledged humbly that we can learn from them, and expressed our desire that they will always be with us? And how might it change our lives to have this rooted deeply within us?

Doing a die-in

Remembering God's love for his animal creation and their place in his plans for redemption strengthens our motivation to be living and acting in ways that do not harm them, but instead enable them to flourish. This is desperately needed today because we are facing unprecedented extinction rates.

April 2019 saw the emergence of Extinction Rebellion (XR), a new climate movement with national groups in 16 countries (at the time of writing). The origins of the group relate to the recognition that we are facing a climate and ecological emergency, as we have discussed at so many points in this book. XR believes that the political and corporate world is not taking the action it should and feels frustrated by the slow rate of change and the lack of urgency in response to this crisis. XR is therefore building a movement that increases pressure on those who could make a difference by means of civil disobedience: it is undertaking acts of peaceful resistance to attract the media's attention and get these issues on the public agenda. Its call is, 'for governments to tell the truth about our emergency; to commit to becoming net zero by 2025, and to create a more participatory democracy in the form of citizen assemblies which will lead on climate and ecological decisions'.[13]

For ten days in April 2019, protesters (called Rebels) gathered in different countries. In London, thousands of people took over five sites: Waterloo Bridge, Piccadilly Circus, Marble Arch, Oxford Circus and Parliament Square. They blockaded the roads with some chaining and gluing themselves to one another and the ground; they held rallies and parties and picnics; they talked to the media and government. Over one thousand people were arrested though the whole demonstration was conducted peaceably (which was not the case in France where the police were more brutal and used tear gas despite the protestors' lack of agression). There have been more mass actions since those days in April and churches, Christians and Christian agencies were and are involved.

I joined in over the mid-point weekend, and my elder daughter Mali stayed up and participated for longer. She took part in a 'die-in' at the Natural History Museum. About a hundred activists – many of them young families with pushchairs – gathered in the iconic main hall and, when a whistle was blown, fell to the floor and lay for 25 minutes in silence, pretending to be dead, until the whistle was blown again and they got up and continued on their way. Performed under the massive skeleton of the blue whale – itself an endangered species – the act was done to symbolize the mass extinction of animals we are currently experiencing, which climate change is making worse.

It is one of five key drivers that are altering nature, the other ones being changes in land and sea use (as we saw in regard to birds and sea creatures); direct exploitation of organisms;[14] pollution, and invasive species. Although currently it is land and sea use (intensive agriculture, urbanization and overfishing) that cause the greatest species decline, it is expected that climate change may become the key driver in some cases.[15] And, of course, they are all interlinked.

Currently, around one in four mammals is at risk of extinction – a terrible statistic.[16] In 2019, the IPBES released its Global Assessment Report on Biodiversity and Ecosystem Services (we saw it briefly in the last chapter in relation to marine fish stocks).[17] The then-IPBES Chair, Sir Robert Watson, said at its launch, 'Ecosystems, species, wild populations, local varieties and breeds of domesticated plants and animals are shrinking, deteriorating or vanishing. The essential, interconnected web of life on Earth is getting smaller and increasingly frayed.' It is also becoming increasingly upside down in that, as we saw with birds in the previous chapter, the vast majority of mammals are livestock.

What was clear from the report is that it is people who are both causing the problems and suffering the consequences, along with other species. From a human perspective, the failure to conserve and use the wider world sustainably is affecting our ability to meet 80 per cent of the Sustainable Development Goals: those related to poverty, hunger, health, water, cities, climate, oceans and land. As the IPBES report says, 'Loss of biodiversity is therefore shown to be not only an environmental issue, but also a developmental, economic, security, social and moral issue as well'.[18] As we have seen throughout *Saying Yes to Life,* people and planet have to be held together.

Underlying the five key drivers of our current levels of extinction and loss of biodiversity is consumerism. Brazilian anthropologist, Professor Eduardo Brondizio, who co-chaired the report, highlights a pattern that emerges: 'one of global interconnectivity . . . with resource extraction and production often occurring in one part of the world to satisfy the needs of distant consumers in other regions'.[19] Our globalized, consumer-focused society has brought us many good things and I am not against consumerism *per se*. The fact that

I can sit and write this book on a laptop, with a cup of tea and some chocolate next to me, comfortably clothed; the fact that I can relax with a variety of different books or television programmes; the fact that I am not spending my days working in the fields to grow food for my family or taking hours to wash our clothes by hand – all these indicate that I am living in and enjoying the benefits of a consumer culture.

And yet we know this culture is causing immense damage and our demands are more than the earth's resources can carry. Paulos Mar Gregorios, Metropolitan of Delhi of the Indian Orthodox Church says, 'In taking what is given by nature, we should be careful to give back to nature what it needs to maintain its own integrity and to supply the needs of the future.'[20] We know we are not doing that.

It sounds obvious to say, but nothing comes from nowhere: everything we buy and use comes from somewhere and uses resources from the land or seas. Put simply, we need to buy and use less in order to take better care of the natural world for its own and for God's sake, and to free up resources for those who truly need them. Ask yourself all the time: Do I need this? Can I do without it? And, if you think you really do need it, is there is a way of buying or using it that uses no resources (e.g. second hand or sharing with someone else) or that uses new resources sustainably? Wherever you live in the world, my guess is that if you are reading this book, you are in a position to take on this challenge, and what better time to do that than now during Lent?[21]

North American poet, Luci Shaw, calls God, 'the original artist of the universe', saying, 'Just as each human thumbprint is unique, its pattern inscribed on the work of our hands and minds, the Creator's is even more so – the original thumbprint on the universe.'[22] From a theological perspective, therefore, the loss of biodiversity is a desecration of God's artistry, worse than any terrorist demolition of a precious world heritage site. A world that God has created to be teeming with life is instead losing its life at an unparalleled rate. Thoughtlessness and selfishness make us all complicit. But – as we shall consider soon – alongside our sinfulness, we also bear the imprint of God in our own lives and that means we can act and bring about hope.

Yasha the pangolin

One animal that has found a special place in our lives as a family is the pangolin. The world's only scaly mammal, it is an incredible creature with eight species across Africa and Asia, some living on the ground and some in trees. The pangolin is covered from head to foot in scales, has an amazingly long tongue for reaching termites, and rolls into a ball to protect itself when threatened. Ground-dwelling pangolins walk on their hind legs in a very endearing manner.

The pangolin is totally harmless but is also the world's most illegally trafficked animal. It is hunted for its meat – both for local bushmeat and for sale around Asia where it is considered a delicacy – and its scales, which are thought to have medicinal properties and are used in traditional medicine, particularly in China (though the scales are made of nothing more than keratin like our nails). All eight species are at risk of extinction and, though globally illegal, over a million pangolins have been trafficked in the last decade alone. Hauls of pangolin scales are regularly found by customs officials. While I was writing this, news came in of Turkish officials finding 1.2 tonnes of scales at Istanbul airport, and the biggest haul ever came in April 2019 when nearly 13 tonnes from Nigeria were seized by customs officials in Singapore, en route to Vietnam. The scales were likely to come from some 17,000 pangolins.[23]

Mali participated in the die-in at the Natural History Museum because she has developed a love of pangolins and wanted to play her part in something to raise awareness around species extinction. She took time out between school and university to volunteer with A Rocha Ghana and be involved in the work they are doing to protect the Atewa Forest. The Atewa Forest, in the south east of Ghana, is 250 square kilometres and the source of three major rivers which together provide water to five million people, as well as supporting livelihoods and agriculture. It is home to over 100 species of mammals, birds, amphibians and plants that are globally threatened or near-threatened with extinction, some of which are found nowhere else in the world. And it is home to pangolins.

The pangolins in the Atewa Forest are hunted for local bushmeat and for the wider international trade. The forest itself is under pressure because large amounts of bauxite – used to make aluminium – are in the ground and the Ghanaian government is working with the Chinese government to open up the forest to extraction. There is also illegal logging and gold mining and land clearance for farming.

Mali's work was focused on going into schools and forest communities to help the local people understand why they should not take pangolins from the forest, and to look at alternative employment so they would not need to sell them as a means of income. Unexpectedly, she rescued a female juvenile pangolin, whom she named Yasha (which means 'saved' in Hebrew), whose tail had been cut off by a chainsaw when an illegal logger was felling a tree. She took the little creature back to her lodgings and – with the advice of a pangolin expert in Zimbabwe – looked after her, nursing her back to health, taking her into the forest twice a day to feed on termites, and teaching her to gain confidence climbing trees, since she did not have her tail to help her balance. Eventually, Mali released the pangolin back into the forest and we pray that Yasha is still an appropriate name for her.

The plight of the pangolin highlights so many issues: poverty, greed, ignorance, deforestation, consumerism, poaching, land use, biodiversity loss, etc. Yet, it also highlights the amazing love and dedication of those working to safeguard them in Africa and Asia, and of those involved in protecting the Atewa Forest. A Rocha Ghana is one of the key organizations engaged in a lengthy battle with the Ghanaian government to keep bauxite mining out of Atewa and protect the area. At the time of writing the Government of Ghana is preparing to offer sections of the forest to mining companies to start extracting bauxite, but local campaigners have not given up hope that these plans can still be averted.

Some of us reading this will be living in countries that have or use pangolins. If that is the case for you, you can help by calling on the government to put more resources into policing and penalizing poaching and by doing what you can to change the culture so it becomes unacceptable to eat or use pangolins. And of course you can make sure you never do so yourself. Others of us can play our part by supporting organizations working for their protection, and for the protection of other endangered animals.

Whoever would have thought that a prehistoric creature called a pangolin would find its way into a Lent book? But it is completely appropriate. For those of us reading *Saying Yes to Life* a chapter a week through Lent, you will be heading towards or maybe even in Holy Week, focusing your thoughts on the events leading up to Jesus' death and resurrection. When all the creatures in heaven and earth in Revelation 5 come together with the elders to worship at the throne of God, they see at the centre the Lion of Judah who has become a slain lamb. Animal imagery is everywhere and animals themselves feature right the way through the biblical story of salvation. With them we worship the one who died so that all creation might be reconciled to God through his blood.

Made in the image of God

Let us return to the text of Genesis 1 that we are looking at in this chapter and consider the last thing God creates: human beings. Immediately we see there is something different here that sets this final species apart from all else that has been created. In verses 11, 20 and 24, the land and seas are told to produce the creatures that will inhabit those spaces. Here though in verse 26, we see 'a special deliberation on the part of God'.[24] The plural that is used ('let us make') may refer to the concept of the celestial court that we see elsewhere in the Old Testament (eg. 1 Kings 22.19–22), and could be God discussing his decision with the other heavenly beings there. Or it may be what French theologian Paul Beauchamp describes as 'the distant dawn of a trinitarian revelation'.[25] Or it may be the 'royal we', an expression of the grandeur and momentousness of the act being undertaken (the Hebrew *'ĕlōhîm*, 'God', is often in the plural, e.g. Isa. 19.4). Whichever of these understandings is correct, the author wants us to understand that the creation of humankind carries particular significance.

We must, however, be careful not to make too much of a distinction. Humans are, after all, created on the same day as the other land creatures. We may have preferred the sea, sky and land animals all to have been made on Day Five with Day Six as our very own special day! But no, we are land creatures along with all the rest of Day Six. Like them, we have the breath of God in us as we have seen already. And we remember what we saw in Chapter Five on the fish and birds, that the blessing given to us to be fruitful and increase and fill the earth (our space) is also given to the sea and sky creatures to fill their spaces (Gen. 1.22). We shall consider further on in this chapter how we should see ourselves in relation to the wider world, but for now it is worth noting the things that make us the same as the other animals created by God, and to allow that to engender humility within us.

Nonetheless, there is one key difference: whereas all the other sea, sky and land animals are made 'according to their kinds', humankind is made instead 'in the image of God', and it is in this description that we learn what makes us uniquely human. There has been much ink spilt over what this phrase means and scholars have tried to locate it in what particular aspect of humankind's being the image of God can be found. S. R. Driver, for example, in his 1904 commentary on Genesis says that 'it can be nothing but the gift of *self-conscious reason*, which is possessed by man, but by no other animal', while Karl Barth believed as God made them male and female, that the image of God referred to human sexuality.[26]

The Hebrew for image, *ṣelem*, is used elsewhere of a physical image or statue of a god (eg. Num. 33.52) and it has its parallels with Israel's neighbours: a Mesopotamian text talks about the king being in 'the image (*ṣalam*) of Bel' and the name Tutankhamun in Egypt means 'the living image of Amon'.[27] The Israelites were therefore well familiar with the concept of a person or object imaging a god. They will also have been familiar with their neighbours putting such images in their temples to signify the presence of their god(s).

In a number of places in the Bible, the world is seen in temple imagery: the cosmos is depicted as a building created to provide a space in which God's creatures can reside (Job 38.4–7), and the building of the physical temple has resonances with the creation of the universe.[28] Canadian Bible scholar, Richard Middleton, thus describes the creation as the 'macrocosmos' and the tabernacle as the 'microcosmos'. Seen from this perspective, just as a temple would be expected to contain statues representing the god/s (or goddess/es) to which it was dedicated, so as human beings we are the images – the representatives – of the one true God in his creation. It is because of this, of course, that the Israelites are commanded not to make any idols or set up any images of God (Lev. 26.1). They should have no need of such things because they themselves are the images of God. To make other images can only mean they have forgotten their calling.

Henri Blocher notices that the term suggests 'beholding'.[29] An image in a temple is something you look at in order to gain a better understanding of the divine. *We* are that image and other people and indeed the whole creation should be able to look at us, individually and as a species, and see God reflected in us.

Being made as the image of God puts humankind in a particular relationship before God. It cannot be a coincidence that, in Genesis 5, after a summary statement which says, 'when God created mankind, he made them in the likeness of God', the author goes straight on to tell us that Adam then had a son 'in his own likeness, in his own image' (Gen. 5.1–2). As Blocher says, 'God creates man as a sort of earthly son, who represents him and responds to him'[30] and of course 'child of God' language is used throughout the Scriptures (e.g. Hos. 11.1; Lk. 11.11–13; Gal. 3.26; Rom. 8.19).

We cannot jump from this to say that other creatures therefore have no relationship with God: we have seen time and time again that the wider creation in its diversity, fullness and individuality responds to God and God responds to and communicates with it in various ways. And we have also observed repeatedly that the natural world reflects God and carries his imprint, so we must be careful in our understanding of people being made to image God – it does not mean that God cannot be seen in any creature other than humans.

Yet, the human being is the species chosen by God to bear his image and be his representative. Being made as God's image-bearers gives us a job to do with which no other part of the created order is tasked. This carries significance in two particular areas: it impacts how we view our relationships with one another, human to human; and it impacts how we view our relationships with the wider creation, and other creatures in particular. Let us look first at what being created as the image of God tells us about human relationships, and then we shall turn to consider our relationship with the rest of creation.

All people made in the image of God

One of the strands running through *Saying Yes to Life* has been an exploration of how seeing the creation story of Genesis 1 in its wider historical context, particularly alongside the dominant narrative of *Enuma Elish*, can help us understand it more and appreciate what it is telling us. This is no less the case when it comes to the creation of humanity.

Right back in Chapter One, when we heard the story of *Enuma Elish*, we saw that humans were created from the blood of the defeated and slain enemy Kingu (Tiamat's consort) to serve the gods and set them free from doing the work that Marduk at first assigned to them (cleaning the temples and arduous jobs like that). Effectively, people were created to be the slaves of the gods – and created out of the blood of an inferior deity. That is hardly a flattering way to see yourself! And then we discovered in the culture of Israel's neighbours that it was the king who was seen as being God's image-bearer: him and no-one else. The rest of the population were, again, slaves. As slaves they had no value and could be used and abused without justification.

It will be immediately obvious how different that is to the creation of humankind in Genesis 1. Created in the image of the supreme creator God, Yahweh, people are endowed with dignity and worth. And not only the king: *all* people. Yes we have been tasked with a job to do, but it is one that gives us responsibility and respect. There is an astounding, radical equality here, and we see the implications of this reverberating through the pages of the Bible, as inequality and oppression are denounced and the people of God are called to live lives that demonstrate justice, mercy and humility (e.g. Ex. 15.1–18; Amos 8.4–7; Micah 6.8; Acts 4.32–37; 1 Cor. 11.17–22; James 2.1–4; 5.1–6).

People are, as the Psalmist puts it, 'fearfully and wonderfully made' (Ps. 139.14). Each one of us has been knit together in our mother's womb, with our innermost being created by God (v. 13). Rowan Williams writes beautifully of how we are each of value before God:

> This means that whenever I face another human being, I face a mystery. There is a level of their life, their existence, where I cannot go and which I cannot control, because it exists in relation to God alone . . . The reverence I owe to every human person is connected with the reverence I owe to God, who brings them into being and keeps them in being. I stand before holy ground when I encounter another person.[31]

Equality does not mean being identical: different skills and different choices will lead to different lives. But, it is this belief in the equality of all people before God that leads Christians to fight so adamantly against poverty, injustice, discrimination and oppression. As we have seen in every chapter, we live in a fallen world – a world where things are not as they should be because of the consequences of sin. When we do not follow the God, whom Jesus reveals, and his ways of love, compassion and faithful service, the results are devastating. They lead to a caste system of deep inequality; to people working in horrendous conditions to provide the goods we want; to children being trafficked for the sex industry or domestic slavery; to colonialism that destroys whole peoples and ways of life; to war and ethnic conflict that wreck many innocent lives, and much more besides. Every time something happens to a person that goes against the value and worth God places on them, 'some unique and unrepeatable aspect of God's purposes has been allowed to vanish.'[32]

This is what drives Tearfund and all our amazing partners and supporters around the world to work to see people freed from poverty and living transformed lives, helped to reach their God-given potential. Christian Aid similarly is committed to working 'for dignity, equality and justice' and CAFOD reaches out to help people living in poverty and campaigns for global justice. We are just a tiny part of a huge global Church with churches, individual Christians and organizations responding practically to people's needs, standing up against oppression and discrimination, and very often putting their own lives on the line as they do so.

The other fifty percent

When thinking about issues of inequality and injustice, one key area is gender. Though Barth may have been wrong in seeing the image of God as located in human sexuality *per se*, he highlighted something that is important: the image of God is found in human beings as male and female together, in equality. For true human flourishing to occur, both women and men need to be able to live their lives to the full. Gender equality is needed because limited education means lower skill levels, leading to fewer opportunities for work and therefore less income. However, increased rates of female education lead to greater social development and enhanced communities. Gender equality is also the best way to tackle overpopulation because, by educating young women, families are started later and the length of time between each child increases. Though there has been good progress, we still face a situation where only 52 per cent of women in marriage or partnership make their own free decisions about sex, contraception and health care; globally, national parliaments are only 37 per cent women; at least 200 million women and girls in 30 countries have undergone Female Genital Mutilation (FGM), and women in northern Africa hold fewer than one in five paid jobs in the non-agricultural sector. Thinking back to Chapter Three on the importance of land rights, globally women own just 13 per cent of agricultural land though they are the majority farmers.[33]

There is a lot of focus on empowering women, in line with this being one of the Sustainable Development Goals (no. 5: Gender equality). Tearfund has seen how cultural and religious norms that justify male dominance and violent behaviour are a significant contributor to lack of female empowerment and so, if we want to see gender equality, we need to tackle those norms.

Mary from the Democratic Republic of Congo (DRC) suffered in an abusive and controlling marriage for more than six years. 'He would come back home drunk at around 10pm, and would immediately attack and insult me in front of the children,' she remembers. Mary is one of a staggering number of women affected by sexual and gender-based violence in the DRC.

Recognizing that faith leaders are in a unique position to speak out against such harmful social norms, Tearfund has pioneered an approach to tackling violence against women that is called 'Transforming Masculinities'. This uses scriptural reflections to change thinking and promote respectful relationships between men and women, and religious leaders of various faiths are being equipped and mobilized to tackle violence against women in their communities.

When Mary's husband started attending the Bible-based Transforming Masculinities sessions at his church, his whole mindset started to alter, and his behaviour followed suit. 'I was so amazed by the way my husband started changing little by little, and acting differently,' says Mary. 'Now, my husband speaks to me in a soft voice, we even get to discuss questions in our home. He comes back at 7pm at the latest, and he is now concerned with the education of the children and, most importantly, not being drunk. I do not know how to thank you enough.'

So far nearly 800 faith leaders have been trained through the workshops and 6,000 men and women have completed the six-week process. The programme is now being run in Liberia, Nigeria, Brazil, Myanmar, DRC, Central African Republic, Iraq and Burundi, with adapted programmes in Chad, Mali and Sierra Leone which focus on FGM and preventing child marriage. As with Mary's husband, this is resulting in amazing changes in behaviour, including decreased violence. It has also led to increased confidence for many more women like Mary, and safer and more positive home lives for children, all of which will lead to greater empowerment for women.[34]

Sharing our common home

So the image of God in which we are made places us in a particular relationship with God and it affirms the equality of all people, women and men together (and of course the wider biblical witness, particularly in the New Testament, would affirm equality across ethnicity too, though that is not the focus of the Genesis text itself, eg. Eph. 2.11–22; Gal. 3.28). Yet it also has something to say about our relationship with the wider natural world and other creatures.

In his encyclical letter, *Laudato Si'*, Pope Francis coined a new term, referring to this world as 'our common home' and calling on every person living on this planet to care for it. He described this common home as being 'like a sister with whom we share our life and a beautiful mother who opens her arms to embrace us', and goes on to say, 'everything is related, and we human beings are united as brothers and sisters on a wonderful pilgrimage, woven together by the love God has for each of his creatures and which also unites us in fond affection with brother sun, sister moon, brother river and mother earth.'[35] In referring to aspects of the natural world in this way, he is drawing on St Francis and his 'Canticle of the Creatures' in which he gives praise to God for brother sun and sister moon, brothers wind and air, sister water and brother fire, mother earth and even sister death.

This way of seeing our relationship with the wider natural world also has resonances with indigenous spirituality. Stan McKay explains,

> Indigenous spirituality around the world is centred on the notion of relationship to the whole creation. We call the earth our mother and the animals are our brothers and sisters. Those parts of creation which biologists describe as inanimate we call our relatives. This naming of creation into our family is an imagery of substance, but it is more than that, because it describes a relationship of love and faithfulness between human persons and the creation.[36]

Contemporary society has moved so far from that understanding: we tend to regard the rest of the created order as 'the environment', something separate to us, which is there to be used as we want.

In Genesis 1.26–28 we have seen that God made humankind in his image to be his representative in the wider world, in the same way that a physical image of a god or goddess would be put in the temple, or images of a king would be set up throughout his territory to signal his lordship. By making people in his image, God has given us delegated authority over his creation (*The Message* talks about us being 'responsible for . . .'). British Old Testament scholar, Chris Wright, makes the point that the grammar used in these verses indicates the role that humans have: 'Because God intended this last-created species, the human species, to exercise dominion over the rest of his creatures, for that reason God expressly and purposefully creates this species alone in his own image'. The sense of the verses could then be read, 'Let us make human beings in our own image and likeness, *so that* they may look after the rest of creation'.[37]

The image of God, therefore, is not so much any innate quality within us but more like a job title. All those qualities that people have tried to single out as being what constitutes the image of God in us – our ability to reason, to love and form relationships, our creativity and so forth – have been given to us by God, not because we are special but because we need those abilities if we are to carry out well the role that God has assigned to us. We recognize, humbly, that other creatures have those qualities too, but we have been blessed to have been given them in special measure.

The idea of 'having dominion/ruling over' and 'subduing' has of course been used to legitimize all sorts of abusive behaviour towards the land, seas and skies and their inhabitants. In each chapter so far we have sadly had to acknowledge examples of such behaviour. But the text itself does not allow such interpretation. 'Subdue' can imply force but, when used of the land (eg. Num. 32.22) is more akin to 'occupy' and is tied to the idea of filling the earth rather than to a sense of brute force. It finds further meaning in Genesis 2.15 where the human is put in the Garden of Eden 'to work it and take care of it'. Subdue in this wider context would then be associated with agriculture. 'Dominion' or 'rule' is itself a neutral word though it often has negative associations because, regrettably, most rulers carried out their role with violence (e.g. Ez. 34.4).[38] But, God expects his rulers to be different, to be servant rulers who exercise their dominion with love and compassion, working for justice and against oppression (eg. Prov. 31.4–9). American Rabbi, David Sears, concludes therefore that 'the divine mandate for man to dominate the natural world is a sacred trust, not a carte blanche for destructiveness'.[39]

So we occupy this special role as caretakers or gardeners of this world. And, against the Babylonion narrative which fashioned human beings out of defeated divine blood, the Genesis 2 creation story says we are formed out of the dust of the ground (v. 7). There is a play on words throughout these opening chapters: the Hebrew word for earth or ground is *'ădāmâ* and the name for our species is 'the *'ādām* (until Eve is created Adam is not a proper name, it is always 'the adam', more like a description). We are literally made from the earth – humans from the humus. We are earth creatures, 'earthy ones', and this signals the integral relationship we have with the world around us.

Genesis 1.26–28 is only one expression of our relationship with the wider world, though it has tended to be the most dominant one, and it is reflected in Psalm 8 (which is picked up by the author of Hebrews in 2.5–8). Yet we have seen clearly throughout *Saying Yes to Life* that other passages, particularly in the Psalms and the Prophets, do not separate us out in the same way but rather place us on an equal footing with the wider world. Richard Bauckham has captured this in his simple phrase, 'the community of creation', and, as we take on the job of looking after what God has made, we do so remembering that we are indeed part of this wider community.[40]

One part of being in this community of creation is that we are not immune to creation's sufferings. Climate change, species loss, plastic pollution, war and poverty all impact people, other creatures, and ecosystems together. It is a tragic reality too that environmental problems impact the poorest the most and, in countries of mixed colour (such as the US), environmental hazards follow race lines.[41]

We cannot tackle poverty without thinking about the air people breathe, the land they live on and the waters they fish in; and we cannot tackle environmental breakdown without thinking about the people living in specific places who contribute to localized problems through their poverty, and wealthy people like ourselves who are causing global problems through our consumerism. That is why Tearfund, and many other organizations, have made it their priority to help people lift themselves out of poverty in a way that enables the natural world to flourish at the same time. In this area, our work focuses on waste management, renewable energy and climate-smart agriculture, all of which are designed to create green jobs and livelihoods, restore the environment and reduce local inequality. We do this alongside advocacy work that calls on governments and businesses to put into place policies and practices that function in favour of those in poverty and the environment, and help wealthier consumers take action to reduce their levels of consumption. You can see more on this from various organizations at <www.spckpublishing.co.uk/saying-yes-resources>.

We are part of a vibrant, wonderful community of creation and there is much to celebrate in it. But perhaps it would be appropriate to finish this section with an awareness that there is also a place for lament. Patriarch Bartholomew says, 'To commit a crime against the natural world is a sin against ourselves and a sin against God'.[42] Lent is a season when we reflect on suffering; sharing in the suffering of Christ; and acknowledging where our sin has resulted in suffering for others and ourselves. As we reach Holy Week, let us pause to repent of the sins we have committed against God's world and his creatures, whether through negligence, weakness or our own deliberate fault.

Holy eating

As part of the community of creation, we also share common food, and green plants are given for food to all the creatures (Gen. 1.29–30). Food is a really important part of our humanity. We use it not only to give us the energy we need but also to build community and personal relationships. For many of us, food is a way by which we express love and nurture. As Canadian theologian, Norman Wirzba, puts it, 'Food is a gift to be gratefully received and generously shared'.[43] Not having enough food to eat or give to our loved ones is a disaster physically and it robs us of much of what it means to be human. The central act of the Christian faith is the commemoration of a meal. Every time we take communion we remember the Last Supper that Jesus shared with his disciples, with its symbols of his broken body and shed blood, and you may well be about to commemorate that meal yourself with your church on Maundy Thursday.

Yet it has become blatantly clear in *Saying Yes to Life* that food today is one of the key drivers behind many of the problems we are facing, through its rearing, growing, transportation, packaging and disposal. With regards to climate change, agriculture is a significant contributor through greenhouse gas emissions from livestock; the cutting down and burning of forests for land; emissions from the production of the chemical inputs used to grow crops (much of which then goes to feed livestock), and emissions from transportation and the production and disposal of packaging. Land use is hugely important for tackling climate change, through soil and trees absorbing carbon (which reduces when land is degraded or, of course, when trees are cut down).

We certainly cannot put the blame wholly on farmers: the vast majority care deeply about the land under their care and we owe them immense gratitude for the work they do. We must remember too that globally 70 per cent of food is produced by smallholder farmers, the majority of whom are women. Behind them, though, is a system regulated by a few huge multinationals (the ABCD group – ADM, Bunge, Cargill and Louis Dreyfus – control 75–90 per cent of the global grain trade)[44], and the overriding aim is to provide us with food that we can buy as cheaply as possible while making the greatest potential profit.

Rabbis Yonatan Neril and Yedidya Sinclair say there are four questions we should ask ourselves in order to eat in holiness. First, *Why am I eating?* Let us ensure we are eating out of healthy desire rather than negative emotional cravings. Second, *How fast do I eat my food?* This urges us to eat our food slowly and consciously, leading to gratitude for what we have. Third, *Where do I eat?* This is an encouragement to eat at a table rather than at our desk or in front of the television. And then fourth, *With whom do I eat?* Let us share our food with others, speaking words of blessing to one another as we do so.[45]

A fifth question we should add is, *Where has my food come from?* We need to increase our understanding of the food we eat: where it has been grown or reared; whether land and trees have needed to be cleared to provide it; whether harmful chemicals have been used; how far it has travelled; whether it is wrapped in plastic, and so on. I have written in more detail on these issues and what we can do practically in both *Just Living* and *L is for Lifestyle*, and there are more resources on the online resources page.

Genesis 1.29–30 makes clear that the original vision for creation was a vegetarian one, for all creatures. Does that mean we should be vegetarian now? To answer that fully needs more space than we have here and I recommend you look at the online resources for helpful links to explore this further.[46] The way I have come to see it is that the 'book ends' of the Bible do not envisage meat-eating (i.e. at the beginning in Genesis 1 and then the vision of the future 'peaceable kingdom' we discussed in Chapter Four, reflecting Isaiah 11.6–9). In between, meat eating is allowed, though it would appear to be as a concession to humanity's sin rather than a positive thing (Gen. 9.1–3); it seems clear that Jesus ate fish and lamb, and no mention of not eating meat is made in the New Testament (other than in relation to meat offered sacrificially), although of course none of this meat was mass-produced in cages or filled with antibiotics.

So it would seem that meat eating is allowed in this current time. However, as Christians we look back to the harmonious relationship between people and other creatures that was envisaged in Genesis 1, and forward to the transformed peaceful creation. When we remember how much God loves all he has made and that animals are and will be part of God's kingdom, it becomes hard to understand why we would want to kill another creature, however ingrained that might be in our culture. I think it can be argued that there are situations where meat eating is permissible, such as where people depend on hunting for their food (not for pleasure) or in situations of real need and hunger, and we may want to argue that mixed farming systems where animals are reared very well, are beneficial to the health of the wider environment. But as Christians we need to challenge ourselves on the extent of our meat eating much more than we currently do. And, if we do choose to eat meat, we must ensure we only do so from animals we know have lived and died well. Personally, I now eat almost no meat (a far cry from the days when I was part of a pig cooperative!) I have stayed flexible for unexpected situations I might find myself in, but don't cook meat at home and am enjoying discovering how delicious vegan food is too. We all need to move towards a more plant-based diet for the sake of the planet.

And it was very good

We have covered big topics in this chapter! As we draw it and Day Six to a close, we see that the divine declaration of goodness changes. Whereas each day finishes with God looking at what he has made and seeing that it is good (except for Day Two), now God looks at *all* he has made and sees that it is very good (Gen. 1.31). I like to think this is an understatement and God surveys all he has made and says, 'Wow, that's fantastic . . . look at this . . . it's amazing!' In fact, we get some glimpse of that in Psalm 104 where the Psalmist says, 'May the Lord rejoice in his works' (v. 31).

I was speaking at a conference recently on the themes of this book, and the day after, a young woman who had gone home overnight came to see me. She had been playing Lego with her young son that morning and had built a house which, she said, she put a lot of time into and was pretty impressive! Her son came over and – wham – knocked the whole thing down and destroyed it. She felt her emotions rising and then, she told me, she sensed God say to her, 'And how do you think I feel about how you treat my home?' It was the first time she had glimpsed God's love for his whole creation and it changed her.

To end this chapter, pray the prayer below and ask God, similarly, to give you a new understanding of how he sees this world and of how you can be his image within it.

For discussion

1 Pope Francis calls this world 'our common home' and Richard Bauckham talks of 'the community of creation'. Reflect on those two terms. What do they say to you?
2 Philip Newell writes, 'The extent to which we fail to reflect the image of God in our lives is the extent to which we have become less than truly human.'[47] What does it mean to you that all people have been made in God's image? Are there people in your circles or society you need to remind yourself have been made in God's image? In what ways do/can you reflect God's image in your own life in relation to other people and the wider world?
3 We have talked a lot in this book about food and the role it plays in many of the issues we have considered. We have looked at

eating fish sustainably; using less plastic in our food; cutting out or reducing our consumption of meat; growing our own food, and supporting farmers who look after their land and don't use lots of chemical inputs. How willing are you to change the way you eat? What will you commit to doing?

4 The final interview features Archbishop Justin. As *Saying Yes to Life* draws to a close and we head into the Conclusion, watch it here (<www.spckpublishing.co.uk/saying-yes-resources>) and consider: what have you learnt through the book? What has been most memorable? Are there things for which you need to repent? What changes are you making as a result?

A prayer from France

Lord be praised for the immensity and the beauty of your creation.
I pray with humility to be every single day more aware of the variety of species on earth and to seek for their protection.

I thank you for this calling to take care of our planet that you put in many hearts, and I pray many others will follow.

I ask you the grace of being able to see the world with your eyes and to always be amazed by the places I'm blessed to go.
In this time of Lent, Lord help me to discern what specific choice I can make to reduce my ecological imprint on the earth and how I can be an encouragement to people around me to think and act about it.

At times when I can be discouraged by the amount of ecological issues, help me to remember I stand before holy ground when I encounter another person and to believe that you can make everything possible.

Prisca Liotard is a French Catholic with a heart for the unity of Christians. She was part of the Community of St Anselm in 2017–2018 where her sisters and brothers gave her the 'Environment' award, a cause that truly matters to her!

Conclusion
The seventh day
(Genesis 2.1–3)

> Thus the heavens and the earth were completed in their vast array. [2]By the seventh day God had finished the work he had been doing; so on the seventh day he rested from all his work. [3]Then God blessed the seventh day and made it holy, because on it he rested from all the work of creating he had done.

As we come to the end of the first story of creation, so too we approach the end of our Lenten travels, turning our eyes now to the empty tomb and the resurrection of Jesus.

We have witnessed a beautiful symmetry in the narrative we have been reading, with its creation of spaces first and then the creatures to inhabit them, and this verse rounds off that narrative, bringing creation to its completion. God now stops and rests. In Exodus 31.17 we are given a different version: 'he rested and was refreshed'; literally, 'he rested and took breath'. There is a sense of enormous fulfilment, of having completed something wonderfully good and breathing a sigh of satisfaction.

We are reminded of the rhythm of day and night and sacred times that we saw on Day Four, as the seventh day is made holy and the Sabbath is confirmed. Seven-day patterns were not unique to the Israelites and appear also in Assyrian and Babylonian writings, but in the Genesis text, that pattern is rooted firmly in the God who has created the heavens and the earth as his temple, and now takes up residence through his people.[1]

On Day Seven, there is no longer the formula that has accompanied each day – 'and there was evening, and there was morning . . .' – and there is the implication that Day Seven does not finish but continues on. And yet we know we do not live fully in Sabbath rest. Tragically, in the very next chapter of Genesis, we see humanity fall from our intended state of *shalom* to a place of discord and enmity on all levels: with God, with one another, and with the wider created order. Remembering the word play around *ʾādām* that we saw in Chapter Six, in Genesis 3.17 we read that the *ʾădāmâ* ('the ground') is cursed because of the *ʾādām*. The rest of the Bible is the story of how God works to bring restoration: to put back to rights what has gone wrong and bring about the Sabbath rest that has been promised.

As we come out of Lent and into Easter Sunday, we proclaim again our belief that Christ has died and Christ has risen – and that he *will* come again. As Amy Plantinga Pauw says, 'Easter is God's seal that the last word on creaturely life will be peace and praise, and the joy of that hope is already seeping into the present.'[2]

As followers of the risen Messiah, we live in the 'overlapping of the ages'. We have the first fruits of the Spirit, like a seal or deposit that guarantees our future inheritance (Rom. 8.23; Eph. 1.13–14), but we are still awaiting that final time. For all its beauty and wonder, we know we inhabit a world of terrible sadness and suffering and we will not escape that while we live in what Ecuadorian theologian, René Padilla, has called 'between the times'.[3] This is a world of wounds and it can be all too easy to bury our heads in the sand, focus on our own lives and refuse to engage in the issues we have touched on in this book, particularly where they require us to make changes, personally and in our churches and broader society.

There is a tension between where we are now and where we look forward to being, described beautifully by British Catholic theologian, Peter Hocken: 'The Spirit has been given both as the first fruits and the hope of full liberation, and we are stretched between the two.' I feel that stretch and it can be painful and difficult. But we know that, as followers of the risen Jesus, we are called to navigate that tension and live lives that speak of his hope for creation. We do that symbolically as we meet each week to pray, worship and break bread together – the Sabbath now not held on the last day of the week, but on the first day of the new week, the resurrection day. And we do that by refusing to give up, remembering that no act of ours is in vain even if we can feel overwhelmed by the tragedies around us.

Resurrection churches

In Tchirozérine in the Republic of the Niger, Pastor Koupra has been helping his church, Coopération Évangélique du Niger (CEN), to consider through a series of Bible studies what it means to care for creation in their local area, and how they can take action as a community. Situated in a desert with little vegetation, the congregation decided that planting trees was a priority and set about digging in hundreds of Neem trees around the community's schools, homes and the church. At the same time, the political crisis was unfolding in neighbouring Libya, resulting in an influx of predominantly Muslim Libyan migrants to the community, who had fled with nothing and had no means of supporting themselves in a new country. CEN made a plan to combine helping the migrants with the work they had been doing to look after the natural world around them. In exchange for food supplies, the migrants supported the tree-planting initiative, as well as helping clear the plastic rubbish in the area, which they used to turn into bricks for building projects. The scheme now supports 650 families in Tchirozérine and nearby Agadez, and has made a visible difference to the natural environment of both cities.

For many years I have carried in my mind an image of the Church as being like a sleeping giant with regards to caring for the whole community of creation. That is not to disregard those who *have* understood the creation-wide implications of the gospel (we could cite, for example, Clement of Rome, Basil the Great, Hildegard of Bingen and Martin Luther, plus others). But on the whole, and particularly in later centuries, the Church has been guilty of moving away from a sacramental and connected relationship with nature to one where nature is simply seen as a resource to be exploited.

We are the largest group of people on the planet – around a third of the global population adheres to the Christian faith.[4] Think what a difference we would make if the sleeping giant awoke and became active!

The good news is that I notice that starting to happen. I see the giant beginning to wake and get out of bed, as all around the world churches are responding to the call to look after the planet entrusted to us. In Thailand, Huay Mai Duei Church is now involved in garbage collection because waste had become a huge problem in the area, and Kha Mu Church has planted a vegetable garden for the local community. In Australia, Tuggeranong Uniting runs a charity shop (known as an 'op shop') to encourage reusing and recycling and they have activists who belong to climate change groups. In the US, Trinity Christian Reformed Church in Michigan promotes caring for the whole creation as an integral part of its preaching and has adopted a stretch of creek that runs near the church to look after. In Argentina, the Church of God in Mendoza has won an award for its litter-picking scheme in the central park, and the Anglican Diocese of Northern Argentina has been monitoring deforestation for the last decade, providing information to the provincial government. In the UK, Portsmouth Cathedral has become the first cathedral to publish its carbon footprint and is actively reducing its emissions, and the Gate Church in Dundee has a whole project dedicated to carbon saving, aiming to become the 'greenest church in Scotland'.

For many churches, prayers and sermons on caring for creation have become a natural part of their church action. Churches have become places where food waste is collected and toddler groups use recycled material that would otherwise have been binned. There are churches going on climate strikes and holding an event with their MP, doing community lunches with plant-based food, and organizing litter picks. There are churches involved in sustainable building projects, zero waste cafés, environmental teaching series, tree planting alongside church planting, toilet twinning, switching to renewable energy and eco-friendly cleaning products, forest schools, using Fairtrade refreshments, installing beehives, moving away from disposable crockery, installing solar panels, holding eco fairs and putting eco-tips in church magazines . . .

This list could go on, though we still have a long way to go and many of us will be in churches that have not yet begun embracing these things and feel discouraged and overwhelmed. Yet, whatever context you are in, I hope these examples will inspire you in your church – whether big or small, urban or rural, of whatever denomination or network – that there are things you can do. How will your church take action to wake the giant?

Resurrection lives

As we wake the sleeping giant of the Church, so we must wake the sleeper inside ourselves too. We now know, biblically, that we are called to look after our common home. And yet, somehow we fail to take serious action in our own lives. When we know the terrible conditions in which the majority of farmed animals are kept, why do we keep buying meat that supports those systems? When we know the immense destruction being done by climate breakdown, why do we refuse to change our flying and travel habits or make a simple decision to eat less meat and dairy? When we know that plastic is causing so much damage to both people and the wider environment, why do we not take easy and obvious steps to use less? When we know our governments and businesses need us to push them to make large-scale changes, why do we stay silent rather than joining our voices with others?

At Tearfund we talk about Pray, Act, Give.

Pray is where we start and what undergirds everything we do. We pray because we believe prayer works and because it changes things – ourselves included. As I write in *L is for Lifestyle*, prayer connects us with the people and situations around the world for whom and for which we are praying. It reminds us of our motivation, which is to see the kingdom of God manifest in our world. It reminds us that there is a strong spiritual dimension to all we do. Above all, prayer reminds us that we cannot do everything by ourselves or in our own strength. Ultimately, we depend on God to bring his redeeming power to bear in the situations we pray for.[5]

As part of our prayer, we must then **Act**. We have considered in the course of this book many difficult, heartbreaking and challenging topics, as well as glimpsing the wonder and diversity of God's creation. If what you have read has made for interesting Lent discussions and nothing else, then it has failed. We are facing a climate crisis, species loss and plastic pollution, and they interweave with a host of other issues, causing conflict, poverty and suffering. We *must* act. We must make bold changes in the way we live – consuming less and consuming better – and take action by pushing our governments and businesses to make bold changes too, including moving to an economic model that enables all people to have what they need, within a flourishing natural environment.[6] As you finish this book and move out of Lent into Easter, go to <www.spckpublishing.co.uk/saying-yes-resources> and look again at the wealth of information that is there. We cannot do everything. But God will break your heart over particular issues, and that is where he is calling you to get involved. What resurrection practices will you take on in your life?

Finally, one very tangible resurrection practice is to **Give**. Giving connects us with people and places around the world as we use our money to bring relief and help change situations. It challenges our own attitude to money and material goods, and causes us to delight in being generous to others rather than focusing on buying more things for ourselves. Of course, we cannot support each and every issue. But, ask yourself today, am I being as generous as I could be? Has God stirred my heart about particular issues in *Saying Yes to Life* that I could start supporting financially?

As we pray, act and give, Day Seven reminds us that we do so as part of a Sabbath rhythm. Yes, the problems are immense and there is much to be done, but our actions must be held within patterns of rest, stillness amid activity.

Writing for the *New York Times* on the moral crisis of climate breakdown and the charge to the Church to respond, Archbishop Justin said, 'As people of faith, we don't just state our beliefs — we live them out. One belief is that we find purpose and joy in loving our neighbors. Another is that we are charged by our creator with taking good care of his creation.'[7]

Resurrection churches, resurrection lives. This is the calling that is on us as we look at all that God has made and say yes to life.

Notes

Introduction

1 My use of the capital for 'Days' is to signal that I do not see these as literal days. There are good questions to be asked around how the Genesis creation texts relate to contemporary science but that is not the focus of this book. To look into this more, I recommend, D. Alexander, *Creation or Evolution: Do we have to choose?* and, E. Lucas, *Can We Believe Genesis Today?* There are also helpful leaflets on the Christians in Science website at <http://www.cis.org.uk/resources/thinking/>.

Chapter 1

1 D. Moo and J. Moo, *Creation Care: A biblical theology of the natural world*, p. 46.
2 Although I have decided to refer to God using male terms, I do so hesitantly and in the full knowledge that God is not male and therefore to use exclusively male terms is unsatisfactory. I find the alternative options clunky and unsatisfactory too, though, so simply ask you to forgive and bear with me if you would have preferred me to have done this differently.
3 W. Maathai, *Replenishing the Earth: Spiritual values for healing ourselves and the world*, p. 22.
4 S. Gitau, *The Environmental Crisis: A challenge for African Christianity*, p. 60.
5 Eriugena, Periphyseon, 841D, in, J. Philip Newell, *The Book of Creation: The practice of Celtic spirituality*, p. 10.
6 J. P. Newell, *Book of Creation*, p. xvii.
7 R. Woodley, *Shalom and the Community of Creation: An indigenous vision*, p. 60.
8 As we saw earlier, *tehom* has its roots in the Akkadian word *tamtu* and the Ugaritic t-h-m, which reflects the Sumerian *tiamat*.
9 T. Ward, *The Celtic Wheel of the Year: Celtic and Christian seasonal prayers*, p. 38.

10 J. Hoyte, *Persistence of Light: in a Japanese Prison Camp, with an Elephant Crossing the Alps, and then in Silicon Valley*.

11 P. Wohlleben, *The Hidden Life of Trees: What They Feel, How They Communicate: Discoveries from a Secret World*, p. 148.

12 Michaela Hau, Martin Wikelski and John Wingfield, 'A neotropical forest bird can measure the slight changes in tropical photoperiod', 1.

13 M. Kolvula, E. Korpimaki, P. Palokangas and J. Viltala, 'Attraction of kestrels to vole scent marks visible in ultraviolet light', 1.

14 <https://ocean.si.edu/ocean-life/fish/bioluminescence>.

15 <https://www.nasa.gov/mission_pages/NPP/news/earth-at-night.html>.

16 <https://earthobservatory.nasa.gov/features/IntotheBlack>.

17 <https://earthobservatory.nasa.gov/features/IntotheBlack>.

18 < https://www.youtube.com/watch?time_continue=12&v=Q3YYwIsMHzw>.

19 L. Worrall and A. Scott, *Pioneering Power: Transforming energy through off-grid renewable electricity in Africa and Asia*, p. 21.

20 Worrall and Scott, *Pioneering Power*, p. 1.

21 Worrall and Scott, *Pioneering Power*, p. 6.

22 IEA, IRENA, UNSD, WB, WHO, 'Tracking SDG 7: The Energy Progress Report 2019'.

23 <https://www.theguardian.com/environment/2015/sep/16/more-people-die-from-air-pollution-than-malaria-and-hivaids-new-study-shows>.

24 Of businesses in sub-Saharan Africa, 50 per cent have identified a lack of reliable electricity access as a major constraint on their business. (Energy Africa Campaign, 2015.)

25 Tearfund runs self-help groups (small-scale community savings schemes) in many of the countries in which it works and they have been incredibly successful and helped to lift literally millions of people out of poverty. Alongside the obvious financial uplift, they offer broad mutual support to members. These groups increase people's resilience to hardship and disaster, particularly the poorest and most marginalized. In Tanzania they are known as Pamoja and you can find out more by just typing Tearfund Tanzania into your search engine.

26 Worrall and Scott, *Pioneering Power*, p. 24.

27 <https://energytransition.org/2016/11/energy-poverty-still-entrenched-in-sa/>.

28 IEA, IRENA, UNSD, WB, WHO, 'Tracking SDG 7: The Energy Progress Report 2019'.

29 International Energy Agency, 'Electricity Information: Overview' <https://webstore.iea.org/download/direct/2261?fileName=Electricity_Information_%202018_Overview.pdf>.

30 L. Worrall and A. Scott, *Pioneering Power.*

31 'Net zero' means that of all greenhouse gas emissions produced/stored/soaked up (e.g by trees or carbon capture and storage), the net result is zero. If there are some emissions that cannot be removed completely, mitigating measures can be taken to ensure they don't go into our atmosphere. In 2019, the UK became the first major economy to commit to a 2050 target in legislation. This was seen as a major step forward though there is pressure to bring the target date forward.

32 IPCC, 'The Special Report on Global Warming of 1.5°C (SR15)'.

33 World Bank Group, 'Shock Waves: Managing the impacts of climate change on poverty', and IPCC 'Special Report'.

34 W. Halapua, *Waves of God's Embrace: Sacred perspectives from the ocean*, pp. 74–76.

35 From a personal conversation.

36 'An interview with the Eco-bishops', *Tearfund Footsteps Magazine*, 99, p. 14.

37 Taken from Katharine's website at <katharinehayhoe.com>.

Chapter 2

1 D. Kidner, *Genesis*, p. 47.

2 Rabbi Yonatan Neril, 'Water: Appreciating a Limited Resource', <https://www.jewishecoseminars.com/water-appreciating-a-limited-resource-longer-article-for-deeper-study/>.

3 <https://www.theguardian.com/environment/2019/mar/18/england-to-run-short-of-water-within-25-years-environment-agency>.

4 While the River Jordan represented liberty for the Israelites in biblical times, today issues of water access in the region, including from the River Jordan, are highly contentious, with Palestinians limited to a fraction of the daily water usage of Israelis (see: <https://www.aljazeera.com/news/2016/06/israel-water-tool-dominate-palestinians-160619062531348.html>). For further information, see also <https://

www.amnestyusa.org/troubled-waters-palestinians-denied-fair-access-to-water/>.

5 T. Wright, *The Way of the Lord*, p. 28.

6 D. Wilkinson, *The Message of Creation*, p. 21.

7 T. Finger, 'An Anabaptist Mennonite Theology of Creation', in C. Redekop (ed), *Creation and the Environment: An Anabaptist perspective on a sustainable world*, p. 164.

8 R. Rohr, *The Universal Christ*, pp. 16–17.

9 T. Finger, 'Anabaptist', p. 164 (Finger mostly uses the masculine in his talk about God, but in this quote he chooses to refer to God in the feminine. He does not give an explanation but presumably this is to show that God is not exclusively male, or maybe this concept is suggestive of pregnancy).

10 Bujo is quoting H. Kessler in Kessler's book, *Das Stöhnen der Natur. Plädoyer für eine Schöpfungsspiritualität und Schöpfungsethik* (Düsseldorf, Patmos Verlag: 1990) and the quote is in, A. Kyomo, 'The Environmental Crisis as a Pastoral Challenge in Africa', in J. Mugambi and M. Vähäkangas, *Christian Theology and Environmental Responsibility*, p. 58.

11 R. Bauckham, *Bible and Ecology*, p. 86.

12 Quoted in R. Valerio, 'Chainsaws, planes, and Komodo dragons: globalisation and the environment', in R. Tiplady, *One Word or Many? The impact of globalisation on mission*, p. 110.

13 A. Kyomo, 'Environmental Crisis', p. 60.

14 R. Rohr, *The Divine Dance: The Trinity and your transformation.*

15 L. Brandt, *Psalms Now*, p. 21.

16 E. Newby, *Slowly Down the Ganges.*

17 <https://www.ancient.eu/Ganges>. There is another tradition that the Ganges was born from the big toe of Vishnu's left foot, and some rivalry between Hindu sects as to which tradition is the correct one.

18 Newby, *Slowly Down the Ganges*, pp. 17–21.

19 R. Cooper, 'Through the Soles of My Feet: A Personal View of Creation', in D. Hallman, *Ecotheology: Voices from South and North*, pp. 211–212.

20 Grooten, M. and Almond, R. E. A. (eds), *WWF. 2018. Living Planet Report – 2018: Aiming Higher*, p. 95.

21 <https://www.metoffice.gov.uk/weather/learn-about/met-office-for-schools/other-content/other-resources/water-cycle>.

22 S. Gitau, *The Environmental Crisis*, p. 129.

23 This and the following paragraphs on salmon come from, P. Wohlleben, *The Secret Network of Nature: The delicate balance of all living things*, pp. 24–27.

24 <https://www.wwf.org.uk/where-we-work/places/ganges> (accessed 16th June 2019).

25 <https://www.iucn.org/news/species/201803/almost-half-madagascar%E2%80%99s-freshwater-species-threatened-%E2%80%93-iucn-report> (accessed 16.06.19).

26 R. Valerio, *L is for Lifestyle*, pp. 163–164. For details see <https://ruthvalerio.net/publications/l-is-for-lifestyle/l-is-for-lifestyle-references/>.

27 Tearfund has been working in this area to help change the situation for Ungwa and her community. For more on this, see <https://learn.tearfund.org/~/media/files/tilz/topics/watsan/wash_new_2018/2018-tearfund-wash-case-study-lulinda-drc-en.pdf>.

28 Water stress occurs when the demand for water exceeds the available amount during a certain period or when poor quality restricts its use. Water stress causes deterioration of fresh water resources in terms of quantity (aquifer over-exploitation, dry rivers, etc.) and quality (eutrophication, organic matter pollution, saline intrusion, etc.) Definition from the European Environment Agency: <https://www.eea.europa.eu/themes/water/glossary>.

29 R. Valerio, *L is for Lifestyle*, pp. 161–162. For details see <https://ruthvalerio.net/publications/l-is-for-lifestyle/l-is-for-lifestyle-references/>.

30 < https://washwatch.org/en/about/about-wash/>.

31 Ocha, G., Hoffman, J. and Tin, T. (eds), *Climate: The force that shapes our world – and the future of life on earth*, p. 132.

32 World Bank, 'High and Dry: Climate Change, Water and the Economy'.

33 Cred crunch, Issue no. 54, April 2019 – Disasters 2018: Year in review United States Agency for International Development; Centre for Research on the Epidemiology of Disasters; Université Catholique de Louvain.

34 <https://www.chiefscientist.qld.gov.au/publications/understanding-floods/flood-consequences>.

35 <http://www.greenanglicans.org/beira-is-the-first-city-to-be-completely-devastated-by-climate-change/>.

36 Thank you to Miles Giljam for this paragraph.

37 United States Agency for International Development; Centre for Research on the Epidemiology of Disasters; Université Catholique de Louvain, 'Cred crunch, Issue no. 54, April 2019 – Disasters 2018: Year in review'.

38 Tearfund is working in Chad to help people like Jumana learn sustainable agricultural practices to help cope with the effects of climate change. To find out more about Tearfund's work in Chad see <https://www.tearfund.org/about_us/what_we_do_and_where/countries/north_and_west_africa/chad/>.

39 To find out more about Toilet Twinning and its approach, see <https://www.toilettwinning.org/our-approach/>.

40 To read more about this work in the DRC, see <https://learn.tearfund.org/~/media/files/tilz/topics/watsan/wash_new_2018/2018-tearfund-wash-case-study-lulinda-drc-en.pdf>.

41 To read more of Dave and Ann's story, see, D. Bookless, *God Doesn't Do Waste: Redeeming the whole of life*.

42 <https://creationtide.files.wordpress.com/2018/07/curry-letter-on-creation_anglican-comm_draft-2-005-final-signed.pdf> (accessed 23rd June 2019).

43 For more on these and other figures see, R. Valerio, *L is for Lifestyle*, p. 166.

44 For more on this see the relevant articles in the Green Living pages of my website <www.ruthvalerio.net>.

Chapter 3

1 <https://www.theguardian.com/environment/2014/jan/20/peru-farmers-weather-climate-change>.

2 Juliana and her husband, Ian, are missionaries with Latin Link. For more on their work in Peru (and a lovely picture of Juliana holding some tree saplings) see <https://www.latinlink.org.uk/peru>.

3 Translation by Dr Hilary Marlow, in personal correspondence.

4 David Wilkinson argues that theological engagement with eschatology has focused too much on the world and neglected the future of the universe as a whole (*Christian Eschatology*, Chapter Three).

5 R. Williams, 'Creation, Creativity, and Creatureliness. The Wisdom of

Finite Existence', in, B. Treanor, B. Benson and N. Wirzba (eds), *Being-in-Creation: Human responsibility in an endangered world*, pp. 25–27.

6 Bauckham, *Bible and Ecology*, p. 15.

7 R. Valerio, *L is for Lifestyle*, p. 20.

8 M. Vähäkangas, 'The Environmental Crisis in the Light of the Cross', in, J. Mugambi and M. Vähäkangas, *Christian Theology and Environmental Responsibility*, pp. 112–113.

9 L. Hart, 'The Earth is a Song Made Visible', in, C. Redekop, *Creation and the Environment*, p. 170.

10 D. Bonhoeffer, *Letters and Papers from Prison*, 485, in A. Plantinga Pauw, *Church in Ordinary Time: A wisdom ecclesiology*, p. 17.

11 David, *Scripture, Culture, and Agriculture*, p. 46.

12 R. Valerio, *L is for Lifestyle*, pp. 19–20.

13 For more on this see E. Davis, *Scripture*, pp. 48–53.

14 The NIV translates 'mourns' in all these verses as 'dries up' (the Hebrew root *ʾābal* carries both senses).

15 <https://www.thenatureofcities.com/2015/12/02/nature-medicine-for-cities-and-people/>.

16 K. Kime, 'Finding Our Voices with Indigenous Australia'.

17 Jocabed Reina Solano Miselis, '*An Mar Nega* (Our Home)', in Hannah Swithinbank and Emmanuel Murangira (eds), *Jubilee: God's answer to poverty?* (Oxford: Regnum, 2020).

18 J. Miselis, 'In Defense of Life and Harmony', *Boletin Teologico*, 11(2), p. 180.

19 <https://www.redletterchristians.org/13-hopes-for-2013/>. R. Williams in the Foreword to, C. Foster, *Sharing God's Planet: A Christian vision for a sustainable future*, p. viii.

20 J. Miselis, 'In Defense', p. 182.

21 K. Kime, 'Finding Our Voices'.

22 R. Cooper, 'Through the Soles', in D. Hallman, *Ecotheology*, p. 208.

23 E. Conradie, 'African Perspectives on the "Whole Household of God" (Oikos)', in I. Mwase and E. Kamaara, *Theologies of Liberation and Reconstruction*, p. 289.

24 J. Mohawk, 'the Tragedy of Colonization', *Indian Country Today*, January 23 2004, cited in, R. Woodley, *Shalom and the Community of Creation: An indigenous vision*, p. 92.

25 S. Gish, *Desmond Tutu: A biography*, p. 101.

26 O. Mungula, 'Indigenous Land Rights in Honduras', in *Tearfund, 'Footsteps'*, issue 105, p. 6; <https://www.tearfund.org/en/2015/12/paradise_won/>.

27 Oxfam, International Land Coalition, Rights and Resources Initiative (2016) Common Ground. Securing Land Rights and Safeguarding the Earth, and Land Matrix, 'Analytical Report of the Land Matrix II: International Land Deals for Agriculture' (2016).

28 J. Evans, *God's Trees: Trees, forests and woods in the Bible*, p. 123. His book has been extremely helpful for the following paragraphs.

29 J. Evans, *God's Trees*, p. 101.

30 L. Brandt, *Psalms/Now*, p. 13.

31 R. Valerio, *Just Living: Faith and community in an age of consumerism*, pp. 269–270.

32 For all of these, Julian Evans gives wonderful background information about the trees themselves to help us understand Jesus' teaching more in *God's Trees*, Chapter 9.

33 J. Evans, *God's Trees*, p. 143 (citing Evelyn's 1664 Silva).

34 T. Ward, *The Celtic Wheel of the Year*, p. 194.

35 P. Wohlleben, *The Hidden Life of Trees: What They Feel, How They Communicate: Discoveries from a Secret World.*

36 P. Wohlleben, *The Hidden Life of Trees*, Chapters One and Two.

37 P. Wohlleben, *The Hidden Life of Trees*, p. 48.

38 P. Wohlleben, *The Hidden Life of Trees*, p. 11 and p. 5.

39 <https://www.iucn.org/resources/issues-briefs/deforestation-and-forest-degradation>.

40 <https://wwf.panda.org/our_work/forests/deforestation_fronts2/deforestation_in_the_congo_basin/> and Tyukavina A *et al.*, 'Congo Basin forest loss dominated by increasing smallholder clearing', *Science Advances*, 4(11) (2018).

41 <https://www.bbc.co.uk/news/science-environment-48827490>.

42 <https://www.worldwildlife.org/threats/deforestation-and-forest-degradation> and <https://www.princeton.edu/news/2018/07/03/southeast-asian-forest-loss-much-greater-expected-negative-implications-climate>.

43 Forest Europe, 'State of Europe's Forests 2015 Report'.

44 For more on rewilding see George Monbiot's brilliant book, *Feral: Searching for enchantment on the frontiers of rewilding.*

45 <https://www.nytimes.com/2019/07/25/world/europe/russia-china-siberia-logging.html>. The forests there are allowed to grow back, meaning the damage caused is less than the clearing of tropical rainforests for agricultural purposes. Nonetheless, the mass deforestation is still problematic.

46 Taylor, R. and Streck, C. *The Elusive Impact of the Deforestation-free Supply Chain Movement* (2018).

47 <https://www.tearfund.org/2017/02/planting_trees_of_life/>; <https://www.tearfund.org/en/2018/09/the_untold_story/>.

48 Global Witness (2018), 'At what cost? Irresponsible business and the murder of land and environmental defenders in 2017'.

49 W. Maathai, *Replenishing the Earth*, pp. 78–79.

50 A. Kyomo, 'The Environmental Crisis as a Pastoral Challenge in Africa', in, J. Mugambi and M. Vähäkangas, *Christian Theology*, p. 60.

51 Thank you to the Bishop of the Anglican Diocese of Amazônia, Bishop Marinez Rosa dos Santos Bassotto, for the information in these paragraphs.

52 <https://www.carbonbrief.org/amazon-rainforest-is-taking-up-a-third-less-carbon-than-a-decade-ago>.

53 Church of South India, 'Green Protocol: Guidelines', p. 29.

54 <http://www.greenanglicans.org/>.

55 <https://www.nature.com/immersive/d41586-019-00275-x/index.html?utm_source=Nature+Briefing&utm_campaign=aafdf211cc-briefing-dy-20190130&utm_medium=email&utm_term=0_c9dfd39373-aafdf211cc-43547817>.

56 D. Bonhoeffer, August 12.1943, Letter to Maria von Wedemeyer, in *A Testament to Freedom: The essential writings of Dietrich Bonhoeffer*, G. Kelly and F. Nelson (eds) (Harper Collins, San Fransisco: 1990), p. 512, in, A. Plantinga Pauw, *Church in Ordinary Time*, p. 17.

57 A. Kyomo, 'The Environmental Crisis', p. 61.

58 For information on palm oil see, 'H is for Habitats', in, R. Valerio, *L is for Lifestyle*, pp. 59–65.

59 For more on this and how offsetting works, see <https://www.climatestewards.org/>.

60 Poem by the Revd Francis Simon, quoted in, S. Gitau, *The Environmental Crisis*, pp. 103–104.

Chapter 4

1 For more on the island and how it has been formational to me as I have wrestled with what it means to follow Jesus in a highly consumer culture, see my, *Just Living*, pp. 1–9.

2 D. Kidner, *Genesis*, p. 49.

3 Full text found at <https://www.ancient.eu/article/225/enuma-elish---the-babylonian-epic-of-creation---fu/>.

4 J. Cherian, H. Dharamraj, J. B. Jeyaraj, F. Philip and P. Swarup (eds), *The South Asia Bible Commentary: A one-volume commentary on the whole Bible*, p. 13.

5 R. Williams, 'Creation, Creativity, and Creatureliness. The Wisdom of Finite Existence', in B. Treanor, B. Benson and N. Wirzba (eds), *Being-in-Creation: Human responsibility in an endangered world*, p. 28.

6 A. Carmichael, *Carmina Gadelica, Vol. I & II*.

7 P. Newell, *The Book of Creation*, p. 53.

8 *Carmina Gadelica III*, 307, cited in P. Newell, *The Book of Creation*, p. 54.

9 T. Ward, *The Celtic Wheel of the Year: Celtic and Christian seasonal prayers*, pp. 1–2.

10 You can find resources at both <https://seasonofcreation.org/> and <https://creationtide.com/>.

11 C. Voke, *Creation at Worship: Ecology, creation and Christian worship*, p. 128.

12 These ideas come from S. and S. Hargreaves, *Outdoor Worship: Engage with God in his creation*.

13 This and the following paragraphs are based on an email conversation with astrophysicist Professor David Wilkinson and on <https://en.wikipedia.org/wiki/Milky_Way>, <https://www.bbc.co.uk/programmes/articles/573w4pMNC36FB96vrsVZnfZ/one-family-worlds-apart>, <https://en.wikipedia.org/wiki/Mars>, <https://solarsystem.nasa.gov/planets/saturn/in-depth/> and <https://solarsystem.nasa.gov/planets/jupiter/in-depth/>.

14 The BBC 2019 series, *The Planets*, was a fascinating exploration of the eight planets of our solar system and of how each has become what it is today.

15 Falchi, F. *et al.*, 'The new world atlas of artificial night sky brightness', *Sci. Adv*. 2, e1600377 (2016).

16 <https://myfwc.com/research/wildlife/sea-turtles/threats/artificial-lighting/> (accessed 7 July 2019).
17 <https://royalsocietypublishing.org/doi/10.1098/rspb.2018.0367>.
18 <https://royalsocietypublishing.org/doi/10.1098/rspb.2019.0364> (accessed 7 July 2019).
19 <https://www.bats.org.uk/news/2018/09/new-guidance-on-bats-and-lighting>.
20 Thank you to Malcolm Guite for this reflection: M. Guite, *Waiting on the Word: A poem a day for Advent, Christmas and Epiphany*, pp. 104–107. Richard Bauckham's poem.
21 < http://www.leftbehind.com/archiveViewer.asp?ArchiveID=104>.
22 R. Bauckham, *The Theology of the Book of Revelation*, pp. 7–8.
23 N. T. Wright, *Jesus and the Victory of God*, p. 635.
24 J. Moltmann, *Theology of Hope*, p. 2.
25 W. Maathai, *Replenishing the Earth*, p. 123.
26 One passage that I am not focusing on but is helpful to mention is 1 Thess. 4.13–18, which has sometimes been taken literally and used to support views on the Rapture. However, the phrase 'to meet' (v. 17) was a term used for a delegation of civic leaders to meet a dignitary who was on their way to the city and escort them in. This passage is therefore not about those who are still alive being whisked off to heaven, but is about Jesus' return to his earth (G. Green, *The Letters to the Thessalonians*, p. 226; N. T. Wright, *The Resurrection of the Son of God*, pp. 214–219).
27 Although Isaiah is one book, it is generally thought that only chapters 1–39 come from the prophet Isaiah himself. Chapters 40–55 were likely written/spoken during the exile, and then chapters 56 onwards speak into the time when the people returned from exile.
28 S. Bouma-Prediger, *For the Beauty of the Earth: A Christian vision for creation care*, p. 77.
29 G. Heide thus gives this translation of 2 Pet 3.10–13 which is helpful: 'But the day of the Lord will come like a thief, in which the heavens as we know them will pass from sight with a roar and the order of this world will be refined with intense heat, and the earth and everything in it will be laid bare for judgment. Since all these things are to be refined in this way, what sort of people ought you to be in holy conduct and godliness, anticipating and hastening the day of God, when the heavens will be

refined by burning and the impure order of this world will melt in the intense heat of judgment! But according to his promise we are looking for renewed heavens and a renewed earth, in which righteousness dwells.' Heide, G. Z. 'What is new about the new heaven and the new earth? A theology of creation from Revelation 21 and 2 Peter 3', *Journal of the Evangelical Theological Society*, 40(1), p. 55.

30 N. T. Wright, *The Resurrection of the Son of God*, p. 463. A good explanation of this chapter is also given by Dave Bookless in *Planetwise* (pp. 83–84).

31 In environmental theology – my own included – it has become the accepted understanding that *neos* means something quantitatively new and *kaine* something qualitatively new. However, a closer look would see the two words as almost synonymous: *neos* is used for new covenant in Heb. 12.24 and Col. 3.10, whereas Eph. 4.24 uses *kainos*. And in Mk. 2.22, the new wine (*neos*) goes into new wineskins (*kainos*). The difference seems to be simply stylistic variation. (My heartfelt thanks to personal correspondence with Richard Bauckham for this insight.)

32 David Wilkinson writes about this brilliantly in *Christian Eschatology and the Physical Universe*. For a summary see pp. 111–114, 133–135 and 186–188.

33 See N. T. Wright, *New Heavens and New Earth: The biblical picture of Christian hope*, and Tom Wright, *Surprised by Hope*.

34 David Wilkinson argues that theological engagement with eschatology has focused too much on the world and neglected the future of the universe as a whole (*Christian Eschatology*, Chapter Three).

35 N. T. Wright, *The Resurrection of the Son of God*, p. 258.

36 C. Wright, 'Creation, Gospel, and Mission', in 'Missional Creation Care', Mission Round Table: The Occasional Bulletin of OMF Mission Research, p. 11.

Chapter 5

1 A Rocha is a Christian conservation charity working for the protection and restoration of the natural world. In case you're wondering about all the flights to the Forum, they are offset through Climate Stewards, which does not solve the issue but does at least make a contribution.

2 For those reading this in a country that can access BBC footage online, you can watch the clip at <https://www.bbc.co.uk/programmes/p00381fg>.
3 Quoted in D. Moo and J. Moo, *Creation Care*, p. 53.
4 D. Bookless, *Planetwise*, p. 65.
5 From personal correspondence with Rabbi Jonathan Wittenberg.
6 J. Stott, *The Birds Our Teachers: Biblical lessons from a lifelong bird-watcher*, p. 10.
7 <http://mentalfloss.com/article/78996/15-amazing-facts-about-15-birds>.
8 Statistics from the IUCN (International Union for the Conservation of Nature), the RSPB (Royal Society for the Protection of Birds) and the Scottish Wildlife Trust.
9 <https://www.birdlife.org/worldwide/news/7-birds-you-won%E2%80%99t-believe-are-threatened-extinction>.
10 <https://www.birdlife.org/worldwide/news/7-birds-you-won%E2%80%99t-believe-are-threatened-extinction>.
11 <https://www.rspb.org.uk/birds-and-wildlife/advice/how-you-can-help-birds/where-have-all-the-birds-gone/is-the-number-of-birds-in-decline>.
12 B.-O. Yinon, R. Phillips and R. Milo, 'The biomass distribution on earth', <https://www.pnas.org/content/115/25/6506>.
13 This article looks at conditions in the EU, but in other places of the world it is even worse: <https://www.theguardian.com/environment/2016/apr/24/real-cost-of-roast-chicken-animal-welfare-farms>.
14 R. Valerio, *L is for Lifestyle*, p. 44.
15 For more on this see the chapters, 'F is for Food' and 'O is for Organic', in R .Valerio, *L is for Lifestyle*.
16 For more, see <https://www.nfuonline.com/nfu-online/news/united-by-our-environment-our-food-our-future/>; and thank you to the Revd Dr Mark Betson, Church of England Rural Officer, for his input here.
17 R. Carruthers, *The Highland Notebook; Or, Sketches and Anecdotes*, pp. 231–232.
18 <https://www.abc.net.au/news/2018-09-24/man-poisoned-wedge-tailed-eagles-in-gippsland-jailed/10298426?fbclid=IwAR2dhkIQuQYJwSLeqaBtFqF6V_LybT8LxXt0eGAoknthr2Yq2rvzj3tWgk4>, <https://www.birdlife.org/worldwide/news/34-andean-condors-found-dead-argentina-poisoning-needs-stop>.

19 You can actually track these five birds by going to <https://www.movebank.org/panel_embedded_movebank_webapp> and typing in 'European roller – Timothée Schwartz – Canal du Midi'.

20 <http://ww2.rspb.org.uk/our-work/rspb-news/news/405835-new-report-reveals-25-million-birds-illegally-killed-in-the-mediterranean-every-year> and <https://www.cambridge.org/core/journals/bird-conservation-international/article/illegal-killing-and-taking-of-birds-in-europe-outside-the-mediterranean-assessing-the-scope-and-scale-of-a-complex-issue/DE4D06F3BD4273B94FD3C9621C615A0A>. See also <https://markavery.info/2017/12/15/guest-blog-millions-slaughtered-birds-richa rd-grimmett/>.

21 <https://www.birdlife.org/worldwide/news/7-birds-you-won%E2%80%99t-believe-are-threatened-extinction>.

22 J. Stott, *The Birds Our Teachers*, pp. 94–95.

23 Jocabed Reina Solano Miselis, '*An Mar Nega* (Our Home)', in Hannah Swithinbank and Emmanuel Murangira (eds), *Jubilee: God's Answer to Poverty?* (Oxford, Regnum: 2020).

24 M. Srokosz and R. Watson, *Blue Planet Blue God: The Bible and the sea*, pp. 108–117.

25 The sea also features in the vision of heaven in Revelation (4.6 and 15.2).

26 W. Halapua, *Waves of God's Embrace: Sacred perspectives from the ocean*, pp. 3–4.

27 W. Halapua, *Waves*, p. 11.

28 W. Halapua, *Waves*, p. 41.

29 R. Sluka, 'The Hidden Things of God in the Ocean', *Journal of Ecotheology*, 43.

30 Adapted from <https://scienceandbelief.org/2015/05/28/in-the-eye-of-the-barracuda-beauty-in-the-ocean/#more-3211>.

31 M. Srokosz and R. Sluka, 'Creation Care of the Other 71%', in, C. Bell and R. White, *Creation Care and the Gospel: Reconsidering the mission of the Church*, pp. 225–226.

32 <https://www.un.org/sustainabledevelopment/oceans/>.

33 <https://www.ipbes.net/news/Media-Release-Global-Assessment>.

34 M. Srokosz and R. Sluka, 'Creation Care', p. 229.

35 <https://www.bbc.co.uk/news/world-africa-25660385>.

36 <https://www.worldwildlife.org/species/shark>.

37 <https://www.iucnredlist.org/>.

38 <https://www.worldwildlife.org/species/shark>.

39 <https://www.bbc.co.uk/news/world-asia-47608949>.

40 Learn more about microplastic pollution, including how to participate in citizen science research on nurdles at <www.arocha.org/microplastics-toolbox>.

41 For more on this see Tearfund's report, 'No Time to Waste: Tackling the plastic pollution crisis before it's too late'.

42 <https://creationtide.files.wordpress.com/2018/07/winston-nz-polynesia.pdf>.

43 For more on this inspiring work, see <https://learn.tearfund.org/en/resources/blog/frontline/2018/06/churches_and_communities_join_forces_to_clean_up_polluted_waters_in_brazil/> and <https://www.thetimes.co.uk/article/why-rubbish-theology-can-improve-lives-l7h8ndg3x>.

44 For more see <http://www.christiansurfers.co.uk>.

45 See <https://oceanconservancy.org/trash-free-seas/international-coastal-cleanup/>. In the UK, the Great British Beach Clean is run by the Marine Conservation Society as part of the global movement (and is on the same weekend): <https://www.mcsuk.org/beachwatch/greatbritishbeachclean>.

46 Fauna and Flora International, IDS, Waste Aid and Tearfund, 'No Time to Waste: Tackling the plastic pollution crisis before it's too late'.

47 For more on this, see 'P is for Plastic' in, R. Valerio, *L is for Lifestyle*.

48 <https://www.seafoodwatch.org/> and <https://www.mcsuk.org/goodfishguide/search>.

49 W. Maathai, *Replenishing the Earth*, pp. 185–186 (adapted).

50 W. Maathai, *Replenishing the Earth*, p. 187.

51 Unlike the other prayers in this book, this one is anonymous and is used by Winston Halapua in *Waves of God's Embrace*.

Chapter 6

1 <https://www.theguardian.com/environment/2018/feb/07/hedgehog-numbers-plummet-by-half-in-uk-countryside-since-2000>.

2 K. Greene-McCreight, *I Am With You*, p. 15.

3 From personal correspondence with Rabbi Jonathan Wittenberg.

4 D. Clough, *On Animals I: Systematic Theology*, p. 36. These paragraphs owe a lot to Chapter 2 of his book, particularly pp. 36–43.
5 R. Bauckham, *Living With Other Creatures: Green exegesis and theology*, p. 181.
6 Bauckham, *Living With Other Creatures*, pp. 111–132.
7 Bauckham, *Living With Other Creatures*, p. 117.
8 Bauckham, *Living With Other Creatures*, p. 116, p. 131.
9 K. Gnanakan, *God's World: A Theology of the Environment*, p. 115.
10 This is of course a very complex issue. To look into it more fully, see chapters six and seven in D. Clough, *On Animals I: Systematic theology*.
11 Bauckham, *Living With Other Creatures*, pp. 97–98.
12 R. Woodley, *Shalom and the Community of Creation: An indigenous Vision*, p. 55.
13 < https://rebellion.earth/> and <https://xrebellion.org/>.
14 This is a technical term that covers a range of things: extraction or harvesting of wild organisms or populations (e.g. by collecting, including fire wood, medicinal or ritual organisms, hunting, fishing etc.); extraction or harvesting of other biological products from an ecosystem (e.g. honey, waxes, etc.); extraction or abstraction of water from aquatic ecosystems (e.g. water withdrawal from streams for food and sanitation, irrigation, inter-basin water transfer), and extraction or abstraction of soils and substrates (e.g. peat).
15 <https://www.ipbes.net/news/Media-Release-Global-Assessment>.
16 <https://www.iucnredlist.org/>.
17 <https://www.ipbes.net/news/Media-Release-Global-Assessment>.
18 <https://www.ipbes.net/news/Media-Release-Global-Assessment>.
19 <https://www.ipbes.net/news/Media-Release-Global-Assessment>.
20 Paulos Mar Gregorios, 'New Testament Foundations for Understanding the Creation', at <https://www.religion-online.org/article/new-testament-foundations-for-understanding-the-creation/>.
21 For a deeper discussion on consumerism and how we live well in our consumer culture as followers of Jesus, see R. Valerio, *Just Living*.
22 L. Shaw, *Thumbprint in the Clay: Divine marks of beauty, order and grace*, p. 21 and p. 26. (As well as an acclaimed poet, Luci is also my aunt, married to my Uncle John whom we met in Chapter One.)
23 <https://www.straitstimes.com/singapore/environment/record-haul-of-

pangolin-scales-worth-52-million-seized-from-container-at-pasir>.

24 S. R. Driver, *The Book of Genesis*, p. 14.

25 H. Blocher, *In the Beginning*, p. 70.

26 S. R. Driver, *Genesis*, p. 15; H. Blocher, *In the Beginning*, p. 81.

27 H. Blocher, *In the Beginning*, pp. 86–87.

28 For a more detailed discussion, see J. Richard Middleton, 'The Ancient Universe and the Cosmic Temple', <https://biologos.org/articles/series/evolution-and-biblical-faith-reflections-by-theologian-j-richard-middleton/the-ancient-universe-and-the-cosmic-temple>.

29 H. Blocher, *In the Beginning*, p. 85.

30 H. Blocher, *In the Beginning*, p. 89.

31 R. Williams, *Being Disciples: Essentials of the Christian life*, p. 64.

32 R. Williams, *Being Disciples*, p. 65.

33 <https://www.un.org/sustainabledevelopment/gender-equality/>.

34 <https://www.tearfund.org/en/2017/11/transforming_masculinities_transforming_lives/>.

35 *Encyclical Letter Laudato Si' of the Holy Father Francis On Care for Our Common Home*, 3 and 68.

36 S. McKay, 'An Aboriginal Perspective on the Integrity of Creation', in D. Hallman, *Ecotheology: Voices from South and North*, p. 214.

37 R. Valerio, *L is for Lifestyle*, p. 22. The Chris Wright quote is from, C. J. H. Wright, *Living as the People of God: The Relevance of Old Testament Ethics*, pp. 82–83.

38 R. Bauckham, *Bible and Ecology*, pp. 16–18.

39 <https://www.jewishecoseminars.com/genesis-and-human-stewardship-of-the-earth-2/>.

40 R. Bauckham, *Bible and Ecology*.

41 Rusty Prichard, 'Mapping Environmental Injustice': a talk given at the 2018 Q Conference, session 6. You can watch this brilliant talk here: <https://q2018.squarespace.com/session-six/#itemId=5af5b792758d468aa505e45f>.

42 Cited in *Laudato Si'*, p. 8.

43 N. Wirzba, *Food and Faith: A theology of eating*, p. 12.

44 <https://blogs.oxfam.org/en/blogs/12-08-03-abcds-global-grain-trade/>.

45 <https://www.jewishecoseminars.com/wp-content/uploads/2018/01/Article-we-are-how-we-eat-a-jewish-approach-to-sustainibilty-LONGER-ARTICLE.pdf>.

46 For an excellent and detailed discussion on this whole topic, see Chapter 2, 'Using Other Animals for Food', in D. Clough, *On Animals II: Theological Ethics*.

47 P. Newell, *The Book of Creation*, p. 84.

Conclusion

1 J. R. Middleton, 'The Ancient Universe and the Cosmic Temple'. Middleton makes the point too that the number seven is associated with the temple in a number of ways, e.g. Solomon took seven years to build the temple and held a seven-day long dedication ceremony.

2 A. Plantinga Pauw, *Church in Ordinary Time*, p. 143.

3 R. Padilla, *Mission Between the Times: Essays on the kingdom*.

4 <https://www.weforum.org/agenda/2019/03/this-is-the-best-and-simplest-world-map-of-religions/>.

5 R. Valerio, *L is for Lifestyle*, p. 8.

6 For more on this, see Tearfund's work around the Restorative Economy: <www.tearfund.org/economy>.

7 <https://www.nytimes.com/2017/11/03/opinion/faith-climate-change-justin-welby.html>.

Bibliography

Books

Alexander, Denis, *Creation or Evolution: Do we have to choose?* (Lion Hudson, Oxford: 2014).

Bauckham, Richard, *The Theology of the Book of Revelation* (Cambridge University Press, Cambridge: 1993).

Bauckham, Richard, *Bible and Ecology: Rediscovering the community of creation* (DLT, London: 2010).

Bauckham, Richard, *The Bible in the Contemporary World* (SPCK, London: 2016).

Bel, C. and White, R., *Creation Care and the Gospel: Reconsidering the mission of the Church* (Hendrickson, Peabody: 2016).

Blocher, Henri, *In the Beginning* (IVP, Downers Grove: 1984).

Bookless, Dave, *Planetwise: Dare to care for God's world* (IVP, Nottingham: 2008).

Bookless, Dave, *God Doesn't Do Waste: Redeeming the whole of life* (IVP, Nottingham: 2010).

Bouma-Prediger, Steven, *For the Beauty of the Earth: A Christian vision for creation care* (Baker Academic, Grand Rapids, MI: 2001).

Brandt, Leslie, *Psalms Now* (Concordia Publishing House, St Louis, MI: 2003).

Carmichael, Alexander, *Carmina Gadelica, Vol. I & II* (1900; repub. Forgotten Books, London: 2007).

Carruthers, Robert, *The Highland Notebook; Or, Sketches and Anecdotes* (A&C Black, Edinburgh: 1843).

Cherian, Jacob, Dharamraj, Havilah, Jeyaraj, Jesudason Baskar, Philip, Finny and Swarup, Paul (eds), *The South Asia Bible Commentary: A one-volume commentary on the whole Bible* (Open Door Publications, Udaipur: 2015).

Clough, David, *On Animals I: Systematic theology* (Bloomsbury T&T Clark, London: 2012).

Clough, David, *On Animals II: Theological ethics* (Bloomsbury T&T Clark, London: 2019).

Davis, Ellen, *Scripture, Culture, and Agriculture: An agrarian reading of the Bible* (Cambridge University Press, Cambridge: 2009).

Driver, Samuel Rolles, *The Book of Genesis* (Methuen & Co., London (rev ed.): 1911).

Evans, Julian, *God's Trees: Trees, forests and woods in the Bible* (Day One Publications, Leominster: 2014).

Foster, C., *Sharing God's Planet: A Christian vision for a sustainable future* (Church House Publishing, London: 2012).

Gish, Steve, *Desmond Tutu: A biography* (Greenwood Press, Westport, VA: 2004).

Gitau, Samson, *The Environmental Crisis: A challenge for African Christianity* (Acton Publishers, Nairobi: 2000).

Gnanakan, Ken, *God's World: A theology of the environment* (SPCK, London: 1999).

Gordon, Robert, *Genesis 1–11 in its Ancient Context* (Grove Books: Cambridge: 2015).

Green, Gene, *The Letters to the Thessalonians* (Eerdmans, Grand Rapids, MI: 2002).

Greene-McCreight, Katherine, *I Am With You: The Archbishop of Canterbury's Lent book 2016* (Bloomsbury, London: 2016).

Grooten, M. and Almond, R. E. A. (eds), *WWF. 2018. Living Planet Report – 2018: Aiming Higher* (WWF, Gland, Switzerland: 2018).

Farrell, Clare, Green, Alison, Knights, Sam and Skeaping, William (eds), *This is Not a Drill: An Extinction Rebellion handbook* (Penguin, London: 2019).

Halapua, Winston, *Waves of God's Embrace: Sacred perspectives from the ocean* (Canterbury Press, Canterbury: 2008).

Hallman, David, *Ecotheology: Voices from South and North* (WIPF & Stock, Eugene, OR: 1994).

Hargreaves, Sara and Hargreaves, Sam, *Outdoor Worship: Engage with God in his creation* (Music and Worship Foundation, Bexley Heath: 2016).

Hoyte, John, *Persistence of Light: in a Japanese Prison Camp, with an Elephant Crossing the Alps, and then in Silicon Valley* (Terra Nova Books, Santa Fe, CA: 2018).

IEA, IRENA, UNSD, WB, WHO, 'Tracking SDG 7: The Energy Progress Report 2019', Washington DC.

Kidner, Derek, *Genesis* (Tyndale Press, London: 1967).

Lucas, Ernst, *Can We Believe Genesis Today?* (IVP, Leicester: 2005).

Maathai, Wangari, *Replenishing the Earth: Spiritual values for healing ourselves and the world* (Doubleday Religion, New York: 2010).

Middleton, Richard and Walsh, B., *Truth is Stranger than it Used to Be: Biblical faith in a postmodern age* (SPCK, London: 1995).

Miselis, Jocabed Reina Solano, 'An Mar Nega (Our Home)', in Swithinbank, Hannah and Murangira, Emmanuel (eds), *Jubilee: God's Answer to poverty?* (Oxford, Regnum: 2020).

Moltmann, Jurgen, *Theology of Hope* (SCM, London: 1960).

Monbiot, George, *Feral: Searching for enchantment on the frontiers of rewilding* (Allen Lane Penguin Press, London: 2003).

Moo, Douglas and Moo, Jonathan, *Creation Care: A biblical theology of the natural world* (Zondervan, Grand Rapids, MI: 2018).

Mugambi, M and Vahakangas, M., *Christian Theology and Environmental Responsibility* (Acton Publishers, Nairobi: 2001).

Mwase, Isaac and Kamaara, Eunice, *Theologies of Liberation and Reconstruction* (Action, Nairobi: 2012).

Newby, Eric, *Slowly Down the Ganges* (Picador, London: 1966).

Newell, Philip, *The Book of Creation: The practice of Celtic spirituality* (Canterbury Press, Canterbury: 1999).

Ocha, George, Hoffman, Jennifer and Tin, Tina (eds), *Climate: The force that shapes our world – and the future of life on earth* (Rodale International, London: 2005).

Padilla, René, *Mission Between the Times: Essays on the kingdom* (Langham Monographs, Carlisle: 2010).

Plantinga Pauw, A., *Church in Ordinary Time: A wisdom ecclesiology* (Eerdmans, Grand Rapids, MI: 2017).

Reddish, Mitchell Glenn, *Revelation* (Smith & Helwys, Macon, GA: 2001).

Redekop, Calvin (ed), *Creation and the Environment: An Anabaptist perspective on a sustainable world* (John Hopkins University Press, Baltimore, MD: 2000).

Rohr, Richard (with Morrell, Mike), *The Divine Dance: The Trinity and your transformation* (SPCK, London: 2016).

Rohr, Richard, *The Universal Christ* (SPCK, London: 2019).

Shaw, Luci, *Thumbprint in the Clay: Divine marks of beauty, order and grace* (IVP Books, Downers Grove: 2016).

Srokosz, M. and Watson, R., *Blue Planet Blue God: The Bible and the sea* (SCM, London: 2017).

Stott, John R. W., *The Birds Our Teachers: Biblical lessons from a lifelong bird-watcher* (Candle Books/Lion Hudson, Abingdon: 1999).

Tiplady, Richard, *One Word or Many? The impact of globalisation on mission* (William Carey Library, Pasadena, CA: 2003).

Treanor, B., Benson, B. and Wirzba, N. (eds), *Being-in-Creation: Human responsibility in an endangered world* (Fordham Press, New York: 2015).

Usher, Graham, *Places of Enchantment: Meeting God in landscapes* (SPCK, London: 2012).

Valerio, Ruth, *Just Living: Faith and community in an age of consumerism* (Hodder Faith, London: 2018).

Valerio, Ruth, *L is for Lifestyle: Christian living that doesn't cost the earth* (IVP, London: 2019).

Voke, Christopher, *Creation at Worship: Ecology, creation and Christian worship* (Paternoster, Milton Keynes: 2009).

Ward, Tess, *The Celtic Wheel of the Year: Celtic and Christian seasonal prayers* (O Books, Winchester: 2007).

Wilkinson, David, *The Message of Creation* (IVP, Leicester: 2002).

Wilkinson, David, *Christian Eschatology and the physical universe* (T&T Clark, London: 2010).

Williams, Rowan, *Being Disciples: Essentials of the Christian life* (SPCK, London: 2016).

Wirzba, Norman, *Food and Faith: A theology of eating* (Cambridge University Press, Cambridge: 2012).

Wohlleben, Peter, *The Hidden Life of Trees: What They Feel, How They Communicate: Discoveries from a secret world* (William Collins, London: 2017).

Wohlleben, Peter, *The Secret Network of Nature: The delicate balance of all living things* (Penguin, London: 2018).

Woodley, Randy, *Shalom and the Community of Creation: An indigenous vision* (Eerdmans, Grand Rapids, MI: 2012).

World Bank, *High and Dry: Climate change, water and the economy* (2016).

Worrall, Leah and Scott, Andrew, *Pioneering Power: Transforming energy through off-grid renewable electricity in Africa and Asia* (Tearfund, Teddington: 2018).

Wright, Chris, *Living as the People of God: The relevance of Old Testament ethics* (IVP, Leicester: 1984).

Wright, Chris, *Old Testament ethics for the people of God* (IVP, Leicester: 2010).

Wright, N. T., *Jesus and the Victory of God* (SPCK, London: 1996).

Wright, N. T., *New Heavens and New Earth: The biblical picture of Christian hope* (Grove Books, Cambridge: 1999).

Wright, N. T., *The Resurrection of the Son of God* (SPCK, London: 2003).

Wright, Tom, *The Way of the Lord* (SPCK, London: 1999).

Wright, Tom, *The New Testament for Everyone* (SPCK, London: 2011).

Wright, Tom, *Surprised by Hope* (SPCK, London: 2011).

Journals and papers

Church of South India, 'Green Protocol: Guidelines'.

Encyclical Letter *Laudato Si'* of the Holy Father Francis On Care for Our Common Home (2015).

Falchi, F. *et al.*, 'The new world atlas of artificial night sky brightness', *Sci. Adv.* 2, e1600377 (2016).

Fauna and Flora International, IDS, Waste Aid and Tearfund, 'No Time to Waste: Tackling the plastic pollution crisis before it's too late' (2019).

Forest Europe, 'State of Europe's Forests 2015 Report'.

Global Witness, 'At what cost? Irresponsible business and the murder of land and environmental defenders in 2017' (2018).

Hau, Michaela., Wikelski, Martin and Wingfield, J., 'A neotropical forest bird can measure the slight changes in tropical photoperiod', 1. *The Royal Society* (1998).

Heide, G. Z., 'What is new about the new heaven and the new earth? A theology of creation from Revelation 21 and 2 Peter 3', *Journal of the Evangelical Theological Society*, 40(1), 1997, pp. 37–55.

International Energy Agency, 'Electricity Information: Overview' (2018). <https://webstore.iea.org/download/direct/2261?fileName=Electricity_Information_%202018_Overview.pdf>.

Kime, Karen, 'Finding Our Voices with Indigenous Australia', Speech given to the Victorian Council of Churches (2016).

Kolvula, Minna, Korpimaki, Erkkl, Palokangas, Palvi and Viltala, Jussi, 'Attraction of kestrels to vole scent marks visible in ultraviolet light'. *Nature* (1995).

Land Matrix, 'Analytical Report of the Land Matrix II: International Land Deals for Agriculture' (2016).

Miselis, 'In Defense of Life and Harmony', *Boletin Teologico*, 11(2) (1982).

'Missional Creation Care', *Mission Round Table: The Occasional Bulletin of OMF Mission Research*, 9(1) (May 2014).

Oxfam, 'Common Ground. Securing Land Rights and Safeguarding the Earth' *International Land Coalition, Rights and Resources Initiative* (Oxfam, Oxford: 2016).

Paulson, G. and Brett, M., 'Five Smooth Stones: Reading the Bible through Aboriginal Eyes' (abridged version of a paper that was to be published in *Colloquium: The Australian and New Zealand Theological Review*) (November 2013).

Sluka, R., 'The Hidden Things of God in the Ocean', *Journal of Ecotheology*, 2 (Spring 2016).

Taylor, Rod and Streck, Charlotte, 'The elusive impact of the deforestation-free supply chain movement' (2018).

Tearfund, 'Natural Resources', *Footsteps 82* (2010).

Tearfund, 'Climate Change', *Footsteps 99* (2016).

Tearfund, 'Land Rights', *Footsteps 105* (2018).

Tyukavina, A. *et al.*, 'Congo Basin forest loss dominated by increasing smallholder clearing', *Science Advances*, 4(11) (2018).

United States Agency for International Development; Centre for Research on the Epidemiology of Disasters; Université Catholique de Louvain, 'Cred crunch, Issue no. 54, April 2019 – Disasters 2018: Year in review'.

Yinon, Bar-On, Phillips, Rob and Milo, Ron, 'The biomass distribution on Earth', *Proceedings of the National Academy and Sciences of the United States*, 115(25) (June 19 2018), pp. 6506–6511 (<https://www.pnas.org/content/115/25/6506>).

Websites

<https://www.abc.net.au/news/2018-09-24/man-poisoned-wedge-tailed-eagles-in-gippsland-jailed/10298426?fbclid=IwAR2dhkIQuQYJwSLeqaBtFqF6V_LybT8LxXt0eGAoknthr2Yq2rvzj3tWgk4 >.

<https://www.ancient.eu/article/225/enuma-elish---the-babylonian-epic-of-creation---fu/>.

<https://www.ancient.eu/Ganges>.

<https://www.arocha.or.ke/ >.

<https://www.bats.org.uk/news/2018/09/new-guidance-on-bats-and-lighting >.

<https://www.bbc.co.uk/news/science-environment-48827490>.

<https://www.bbc.co.uk/news/world-africa-25660385>.

<https://www.bbc.co.uk/news/world-asia-47608949>.

<https://www.bbc.co.uk/programmes/articles/573w4pMNC36FB96vrsVZnfZ/one-family-worlds-apart>.

<https://www.bbc.co.uk/programmes/p00381fg>.

<https://biologos.org/articles/series/evolution-and-biblical-faith-reflections-by-theologian-j-richard-middleton/the-ancient-universe-and-the-cosmic-temple>.

<https://www.birdlife.org/worldwide/news/7-birds-you-won%E2%80%99t-believe-are-threatened-extinction>.

<https://www.birdlife.org/worldwide/news/34-andean-condors-found-dead-argentina-poisoning-needs-stop?fbclid=IwAR0EGcmj4Hsq9HnZj3TJRMjhHD_VQssp86UP7G4euUJ-xZt8DnX_HWjXwMc>.

<https://www.cambridge.org/core/journals/bird-conservation-international/article/illegal-killing-and-taking-of-birds-in-europe-outside-the-mediterranean-assessing-the-scope-and-scale-of-a-complex-issue/DE4D06F3BD4273B94FD3C9621C615A0A>.

<https://www.chiefscientist.qld.gov.au/publications/understanding-floods/flood-consequences>.

<http://www.cis.org.uk/resources/thinking/>.

<https://www.climatestewards.org/>.

<https://creationtide.com/>.

<https://creationtide.files.wordpress.com/2018/07/curry-letter-on-creation_anglican-comm_draft-2-005-final-signed.pdf>.

<https://creationtide.files.wordpress.com/2018/07/winston-nz-polynesia.pdf>.

<https://earthobservatory.nasa.gov/features/IntotheBlack>.

<https://www.eea.europa.eu/themes/water/glossary>.

<http://www.greenanglicans.org/>.

<http://www.greenanglicans.org/beira-is-the-first-city-to-be-completely-devastated-by-climate-change/>.

<http://www.leftbehind.com/archiveViewer.asp?ArchiveID=104>.

<https://www.ipbes.net/news/Media-Release-Global-Assessment>.

<https://www.iucnredlist.org/>.

<https://www.iucn.org/news/species/201803/almost-half-madagascar%E2%80%99s-freshwater-species-threatened-%E2%80%93-iucn-report>.

<https://www.iucn.org/resources/issues-briefs/deforestation-and-forest-degradation>.

<https://www.jewishecoseminars.com/genesis-and-human-stewardship-of-the-earth-2/>.

<https://www.jewishecoseminars.com/water-appreciating-a-limited-resource-longer-article-for-deeper-study/>.

<https://www.jewishecoseminars.com/wp-content/uploads/2018/01/Article-we-are-how-we-eat-a-jewish-approach-to-sustainibilty-LONGER-ARTICLE.pdf>.

<http://katherinehayhoe.com>.

<https://www.latinlink.org.uk/peru>.

<https://markavery.info/2017/12/15/guest-blog-millions-slaughtered-birds-richard-grimmett/>.

<https://www.mcsuk.org/beachwatch/greatbritishbeachclean>.

<https://www.mcsuk.org/goodfishguide/search>.

<http://mentalfloss.com/article/78996/15-amazing-facts-about-15-birds>.

<https://myfwc.com/research/wildlife/sea-turtles/threats/artificial-lighting/>.

<https://www.metoffice.gov.uk/weather/learn-about/met-office-for-schools/other-content/other-resources/water-cycle>.

<https://www.nasa.gov/mission_pages/NPP/news/earth-at-night.html>.

<https://ocean.si.edu/ocean-life/fish/bioluminescence>.

<https://www.nature.com/immersive/d41586-019-00275-x/index.html?utm_source=Nature+Briefing&utm_campaign=aafdf211cc-briefing-dy-20190130&utm_medium=email&utm_term=0_c9dfd39373-aafdf211cc-43547817>.

<https://www.nfuonline.com/nfu-online/news/united-by-our-environment-our-food-our-future/>.

<https://www.nytimes.com/2017/11/03/opinion/faith-climate-change-justin-welby.html>.

<https://oceanconservancy.org/trash-free-seas/international-coastal-cleanup/>.

<https://www.princeton.edu/news/2018/07/03/southeast-asian-forest-loss-much-greater-expected-negative-implications-climate>.

<https://q2018.squarespace.com/sessionsix/#itemId=5af5b792758d468aa505e45f>.

<https://rebellion.earth>.

<https://www.redletterchristians.org/13-hopes-for-2013/>.

<https://www.religion-online.org/article/new-testament-foundations-for-understanding-the-creation/>.

<https://royalsocietypublishing.org/doi/10.1098/rspb.2018.0367>.

<https://royalsocietypublishing.org/doi/10.1098/rspb.2019.0364>.

<https://www.rspb.org.uk/birds-and-wildlife/advice/how-you-can-help-birds/where-have-all-the-birds-gone/is-the-number-of-birds-in-decline>.

<http://ww2.rspb.org.uk/our-work/rspb-news/news/405835-new-report-reveals-25-million-birds-illegally-killed-in-the-mediterranean-every-year>.

<https://ruthvalerio.net/publications/l-is-for-lifestyle/l-is-for-lifestyle-references/>.

<https://www.seafoodwatch.org/>.

<https://scienceandbelief.org/2015/05/28/in-the-eye-of-the-barracuda-beauty-in-the-ocean/#more-3211>.

<https://seasonofcreation.org/>.

<https://solarsystem.nasa.gov/planets/jupiter/in-depth/>.

<https://solarsystem.nasa.gov/planets/saturn/in-depth/>.

<https://www.straitstimes.com/singapore/environment/record-haul-of-pangolin-scales-worth-52-million-seized-from-container-at-pasir>.

<https://tcbmcleish.wordpress.com/2014/11/03/the-20-creation-stories-in-the-bible/>.

<https://www.tearfund.org/about_us/what_we_do_and_where/countries/southern_africa/tanzania/>.

<www.tearfund.org/economy>.

<https://learn.tearfund.org/~/media/files/tilz/topics/watsan/wash_new_2018/2018-tearfund-wash-case-study-lulinda-drc-en.pdf>.

<https://learn.tearfund.org/en/resources/blog/frontline/2018/06/churches_and_communities_join_forces_to_clean_up_polluted_waters_in_brazil/>.

<https://www.tearfund.org/en/2015/12/paradise_won/>.

<https://www.tearfund.org/en/2017/11/transforming_masculinities_transforming_lives/>.

<http://www.theanimalfiles.com/mammals/elephant_shrews/golden_rumped_elephant_shrew.html>.

<https://www.theguardian.com/environment/2016/apr/24/real-cost-of-roast-chicken-animal-welfare-farms>.

<https://www.theguardian.com/environment/2018/feb/07/hedgehog-numbers-plummet-by-half-in-uk-countryside-since-2000>.

<https://www.theguardian.com/environment/2019/mar/18/england-to-run-short-of-water-within-25-years-environment-agency>.

<https://www.thenatureofcities.com/2015/12/02/nature-medicine-for-cities-and-people/>.

<https://www.thetimes.co.uk/article/why-rubbish-theology-can-improve-lives-l7h8ndg3x>.

<https://www.un.org/sustainabledevelopment/gender-equality/>.

<https://www.un.org/sustainabledevelopment/oceans/>.

<https://washwatch.org/en/about/about-wash/>.

<https://en.wikipedia.org/wiki/Mars>.

<https://en.wikipedia.org/wiki/Milky_Way>.

<https://www.worldwildlife.org/threats/deforestation-and-forest-degradation>.

<https://www.worldwildlife.org/species/shark>.

<https://wwf.panda.org/our_work/forests/deforestation_fronts2/deforestation_in_the_congo_basin/>.

<https://www.wwf.org.uk/where-we-work/places/ganges>.

<https://xrebellion.org>.

<https://www.youtube.com/watch?time_continue=12&v=Q3YYwIsMHzw>.